Irrigation Management: Principles and Practices

Irrigation Management: Principles and Practices

Editor

Sharad Sawant

scitus
academics

Irrigation Management: Principles and Practices

Edited by **Sharad Sawant**

ISBN: 978-1-68117-075-6

Library of Congress Control Number: 2015935473

SCITUS Academics LLC,
www.scitusacademics.com
Box No. 4766, 616 Corporate Way,
Suite 2, Valley Cottage,
NY 10989

Printed in United States of America on Acid Free Paper ∞

Contents

Preface

Proper water management planning should take into account all water uses, from the source of irrigation water to the use of plant water. It is therefore very important to distinguish between the requirements of crop water and irrigation or perhaps the requirements of the production system for water. Best management practices (BMPs) for the use of irrigation water can improve consistency and efficiency and reduce water resources contamination. Since every farm is different, producers need to evaluate their systems to determine which BMPs are ideal for their operations.

This book contains cutting-edge information on water use and irrigation management, along with some references to irrigation systems. The soil characteristics that affect the management of irrigation include water intake, water holding capacity and soil erosion. Soil texture, content of organic matter, soil structure and permeability influence these qualities and can limit the management and system options of producers. Therefore, no irrigation system of any type is universally more efficient than another. Producers must know the predominant type of soil that receives irrigation water in each field. The available water holding capacity should be worn for scheduled irrigation with the current depletion status.

The book provides knowledge on the management of drainage and irrigation systems in today's world. Irrigation is an important contribution to agricultural output and overall growth. More than 80% of the water resources widely available are currently utilized for irrigation purposes. Undoubtedly, the modernization of the irrigation system such as concrete lining on the inner surface of the open channel, canal automation, etc. will save water significantly. This book also focuses on the traditional techniques and practices in the area of irrigation management.

Chapter 1

Introduction of Fertigation in Sugarcane Production for Optimization of Water and Fertilizers Use

Daffa Alla Mohamed Abdel Wahab

Agricultural Engineering Section, Research and Development Department, Kenana Sugar Company, Rabak City, Sudan

ABSTRACT

A fertigation experiment was conducted during 2009/2010 and 2010/2011 seasons under commercial field conditions in Kenana Sugar Scheme, Sudan (latitude 13.10′N and longitude 32.40′E) in heavy clay soils, with 65% clay, 24% silt, 11% sand and pH 7.5 - 8.5. The primary objective of this study was to compare different strategies for timing of injection, to develop management practices on the efficient use of water and fertilizers in production of sugarcane, maximize yield, and improve quality. In both seasons four treatments were studied in a randomized complete block design with three replications as:

Injection of the dissolved urea (46% N) during entire irrigation time (100%), during the first half of the irrigation time, during the second half of the irrigation time, and Mechanical application of urea. The first three treatments were applied with the third irrigation cycle through the irrigation water so that nitrogen fertilizer in form of dissolved urea (46% N) was injected in irrigation water by means of pressure differential tank system which is based on the principle of a pressure differential being created by a valve and flow regulator forcing and injecting a varying amount of dissolved fertilizer into the irrigation water. In the fourth treatment the fertilizer spreader and rigid tine cultivator were used for urea application just before the third irrigation event. The reference evapotranspiration (ETo) and the amount of water required for irrigating sugarcane plants were calculated according to its phenological stages using Peman-Monteith approach. The results of this study support the general finding that the injection during the entire irrigation (100% of the irrigation) produced the best distribution uniformity of added urea. In this study, injection during the first half of the irrigation was not statistically different from injection during 100% of the irrigation.

INTRODUCTION

Sugarcane being a giant crop producing huge quantity of biomass generally demands higher amounts of nutrient elements [1] . A large number of research experiments have clearly demonstrated that for producing higher cane and sugar yields on a sustainable basis application of adequate amounts of fertilizer nutrients viz. nitrogen, phosphorous and potassium is essential. Sugarcane is a high bio-mass producer in a sense that it is one of the most photo-synthetically efficient C_4 plants. [2] reported that sugarcane can remove about 300 kg nitrogen/ha/season from the soil (126 Kg itrogen/feddan/season).

[3] stated that the nitrogen requirement of sugarcane at Pongola in South Africa is greatest during the tillering (formative) phase. This is required for adequate tiller production and canopy development. Tillering in field grown sugarcane commences around 30 to 45 days after planting. Therefore, adequate N supply should be made available to the crop in the soil from the start of tillering phase. Further, [4] speculated that the crop requirement for N is higher in early grand

growth period. This facilitates cane formation, checks tiller mortality and promotes cane growth. Application of more nitrogen at later phase of active crop growth period not only promotes late tiller formation, but also affects sugar recovery due to reduced juice sucrose percent, increase in soluble N in juice, water shoot formation besides attracting pests and diseases.

Application of commercial fertilizers is an economic necessity on many soils to promote crop production with nitrogen, phosphorous and potassium being the common fertilizer elements [5] . Nitrogen promotes rapid succulent growth; phosphorus gives early root growth, blooming, and resistance to pest and weather damage. Potassium lends toughness and strength and pest resistance to plants [6].

Significant benefits from fertigation cannot be expected unless the irrigation system design is optimized to meet local soil conditions [7] and this should be the first consideration when appraising fertigation strategies for poor soils.

Fertigation using furrow irrigation is a cost-effective and simple method of applying fertilizer across the field. In the past, fertigation has been associated with micro-irrigation and sprinkler irrigation systems. Injecting nitrogen into irrigation water has become increasingly common for producers using surface irrigation. Potentially, fertigation enables surface irrigation producers a means to more readily attain recommended nitrogen management guidelines aimed at minimizing nitrogen losses from agricultural fields. Moreover, Beth and Filters [8] revealed that fertigation allows growers to apply nutrients in small amounts throughout the season in response to crop needs without the potential crop damage or soil compaction caused by machinery application methods.

The introduction of fertigation technology (combined irrigation and fertilization) in modern agriculture opened up new possibilities for optimal water and nutrient supplies to crops and maintaining the desired concentration and distribution of ions and water in the soil. Fertigation is a key factor in modern intensive irrigated agriculture and can be practices under flood, furrow, and sprinkler irrigation [9] . At the same time, the costs of chemical fertilizers have increased and there is a need to improve fertilizer use efficiency for more benefits. The best answer to this challenge is "Fertigation", where both water and fertilizers are delivered to crop simultaneously through an irrigation system. Fertigation ensures that essential nutrients are supplied precisely at the area of most intensive root activity according to the specific requirements of sugarcane crop and type of soil resulting in higher cane yields and sugar recovery.

Playán and Faci [10] stated that when applying fertilizers through a fertigation system, there are several benefits. Applications can be targeted to specific areas, less equipment is used and fertilizer is applied into the soil, where it will be most effective. This means fertilizer can be applied at a lower rate and be more efficient and inexpensive to suit the exact needs of the plant. With fertigation, compaction due to equipment travel is eliminated. Also, since heavy spreader equipment isn't needed out on the course, equipment costs are lowered drastically. The onetime cost of a fertigation system outweighs the multiple costs of spreading equipment and labour costs heavily. Not having to be out on the course also means that you can apply chemicals in conditions that would make conventional application damaging or impossible.

When fertilizing through fertigation with a well planned and monitored schedule, there is less chance of leachable nutrients not being absorbed by the plant. Usually, the application can be applied with irrigation schedules so that the liquid fertilizer will only penetrate the first two to three inches of the soil. If timed right, deep percolation at application time is avoided meaning less likelihood of leaching.

The chemistry of fertigation is essential factor in the success of the technique. Its scope covers interaction between fertilizers and irrigation water, behaviour in soil solution and rate of availability to plants, influence on soil structure, the compatibility of commercial fertilizers with application techniques, the composition of nutrient solutions and crop response. Most of the available studies rely on methods to assess the appropriateness of different fertigation strategies [11] . For optimization of fertilizers application, the nutrients supply will be adjusted to the specific requirements of the crop in different phonological stages of growth and development. [12] stated that modern fertigation equipment (Fertilizer injection methods) must be used and should be able to regulate quantity of fertilizer applied, duration of applications, proportion of fertilizes, and starting and finishing time.

[13] recommended that the fertilizer have to be injected at a constant rate during the entire irrigation event. This recommendation assumes that the tail water runoff will be blended with other water

and reused in another field. On the other hand, [14] suggested that the fertilizer should be added during the end of the irrigation event to avoid deep penetration of fertilizers to the groundwater. Preliminary studies have indicated that the timing and duration of fertigation applications during the surface irrigation event play a critical role in determining the distribution of fertilizer in the field and the potential movement of nitrate. This is mainly significant in light of evidence that suggests that fertigation may increase rather than decrease leaching of mobile chemicals from crop root zones. [15] conducted a series of fertigation experiments and concluded that the most effective fertigation application practice is expected to be related uniquely to the furrow irrigation method. They added that the infiltration rate of the soil was the most important factor in the choice of injection strategy.

Benefits of Fertigation

Fertigation is a very efficient technique for conserving both water and nitrogen fertilizer and increasing crop production. Because fertigation offers several distinct advantages in comparison to conventional application methods which can be summarized as follow:

¾ Distribution of plant nutrients more evenly throughout the wetted root zone resulting in increased nutrient availability and uptake contributing to higher crop growth rates and cane yields.

¾ Application of nutrients to the soil when crop or soil conditions would otherwise prohibit entry into the field with conventional equipment.

¾ No damage to the crop by root pruning, breakage of leaves, or bending of leaves, as occurs with conventional fertilizer application methods/equipment.

¾ Less energy is expended in application of the fertilizer.

¾ Usually less labour and equipment are required for application of the fertilizer and to supervise the application.

Objectives of the Study

The main objective is to develop management practices on the efficient use of water and fertilizers in production of sugar cane, maximize yield, and improve quality. The specific objectives are to:

1. Compare the conventional fertilization (Figure 1) method with fertigation (Figure 2 and Figure 3).
2. Provide recommendations on irrigation scheduling of gated pipe irrigation for optimal water requirements and irrigation production efficiency plus fertilizer use efficiency.
3. Generation of fertilizers formulas (fertigated) for N, P, and K associated with each method of irrigation for optimal yield and efficient fertilizer use efficiency.
4. Build up experiences and new knowledge on advanced technologies of fertigation.
5. Increase the income of sugar cane producer through reducing the input from fertilizers while increasing crop yield.

MATERIALS AND METHODS

Preface

In this study urea fertilizer (46% N) is applied as a single dose just before the second irrigation. Normally, the

Figure 1: Uneven fertilizer distribution using the conventional fertilization method.

Figure 2: Injection of the dissolved urea fertilizers into the irrigation water by means of 2 inches internal diameters siphon tube.

Figure 3: Injection of the dissolved urea fertilizers into the irrigation water by means of pressure differential tank system.

standard practice adopted under Kenana conditions with regard to application of the nitrogen fertilizer in the form of urea (46% N) which is readily soluble in water is that; it is applied at a rate of 92 kg nitrogen/ feddan (200 kg urea/feddan) in the furrows using locally fabricated fertilizer distributor. The applied fertilizer was then incorporated in the soil using rigid tine cultivator.

The present study has considered nitrogen fertilizer in form of urea mainly because it is subject to a greater range of losses than either phosphorous or potassium, and therefore offers more scope for improvements in management. Urea is readily soluble in water and its solubility ratio is 119 grams per 100 grams water at a temperature 25 degrees Celsius.

Site Description

Location and Climate

A fertigation experiment was conducted during 2009/2010 and 2010/2011 seasons under commercial field conditions in Kenana Sugar Scheme, Sudan. Kenana Scheme is located on the eastern bank of the White Nile river, 300 km south of Khartoum at an altitude of 410 m above mean sea level (msl), latitude 13.10′N and longitude of 32.40′E. The climate of the locality is tropical, dry, hot semi-arid with daily mean maximum temperature of 36.4°C and the corresponding mean minimum temperature was 20.2°C, relative humidity at 08:00 a.m. was about 47% (Table 1(a)). The mean annual effective rainfall as shown in Table 1(b) was 278 mm (for the period 1977 to 2013).

Table 1: (a) Mean monthly reference evapotranspiration (ETo) calculated according to (Penman-Monteith) for the period 1977-2013 in Kenana Scheme*; (b) Mean monthly crop water requirement (CWR) calculated according to (Penman-Monteith) for the period 1977-2013 in Kenana Scheme

(a)

Month	Min. Temp. (°C)	Max Temp (°C)	Relative humidity (%)	Wind speed (Km/day)	Sunshine (h)	Radiation M·Jm−2	ETo Penman-Monteith (mm/day)
Jan	12.6	31.3	38	170	10.2	21.3	5.6
Feb.	17.1	36	35	153	9.5	22	6.3
March	18.7	38.6	25	139	9.8	24.1	7.2
April	21.2	41.4	18	148	10.7	26.1	7.2
May	23.2	41.3	27	114	10.2	25	6.7
June	24	37	56	225	9.1	22.9	6.3
July	23.4	36.2	69	132	7.5	20.7	4.6
Aug.	21.7	31.7	81	225	5.8	18.3	4.2
Sept.	21.8	34.1	73	225	7.8	20.1	4.6
Oct.	21.6	37.8	58	180	10	23	4.9
Nov.	19.6	35.8	38	118	10.5	22	5.4
Dec	17.1	35.4	40	163	10.3	20.8	5.1
Mean	20.2	36.4	46.5	166	9.3	22.2	5.7

(b)

Month	ETo Penman-Monteith (mm/day)	Kc-value	CWR (mm/day)	Days of the month	CWR (mm/month)	Effective rainfall (mm/month)	CWR (mm/month)
Jan	5.6	0.6	3.36	31	104	0.0	104
Feb	6.3	0.7	4.41	28	123	0.0	123
March	7.2	0.8	5.4	31	167	0.0	167
April	7.2	0.8	5.76	30	173	0.0	173
May	6.7	1.1	7.37	31	228	21.0	207
June	6.3	1.3	8.19	30	246	55.0	191
July	4.6	1.2	5.52	31	171	52.7	118
Aug	4.2	1.0	4.2	31	130	90.6	40
Sept	4.6	1.0	4.6	30	138	46.2	92

Oct	4.9	1.0	4.9	31	152	12.5	139
Nov	5.4	1.0	5.4	30	162	0.0	162
Dec	5.1	0.9	4.59	31	142	0.0	142
Total	68.1	11.4	63.7	365	1938	278.0	1660
Mean	5.7	0.9	5.3	30.4	161	23.2	138

*Data taken from monthly weather reports of Sugarcane Research Meteorological Station at Kenana.

The Soil

The soil of the area is heavy montmorillonitic clay, brown in colour, uniform and alkaline in reaction (pH ranged between 7.5 and 8.5). It is non-saline non-sodic containing about 65% clay, 24% silt and 11% sand, with very low infiltration rate (300 $cm^3 \cdot m^{-1} \cdot hr^{-1}$). The mean values of the soil physicochemical parameters are presented in Table 2(a) and Table 2(b).

Treatments

In both seasons four treatments were studied in a randomized complete block design with three replications as; Injection of the dissolved urea (46% N) during entire irrigation time (100%). during the first half of the irrigation time, during the second half of the irrigation time, and Mechanical application of urea using fertilizer distributor and rigid Tyne cultivator. The first three treatments were applied with the third irrigation cycle through the irrigation water.

In the present study the sugarcane cultivar (Co 6806) was planted in plots each containing 20 rows, 650 m. long and 1.5 m between ridges in a randomized complete block design with three replications. The cultivar Co 6806 which was originated in Coimbatore districts of India is considered as high quality cultivar, it performs well in all types of soil. The cultivar Co 6806 is considered as the most popular and leading commercial cultivar. Among the profitable cultivars grown, Co 6806 is now expanded over 80% of the area devoted for sugarcane plantation at the Sudanese sugar estates. It was introduced during the late seventies. In virtue of its high genetic capacity to produce high yields of cane and sugar besides its ability to resist endemic pests and diseases namely smut, it remained to be as number one across sugar estates and represents a major fruitful achievement of the sugarcane introduction strategy during the late seventies.

The cultivar Co 6806 is classified as an early season cultivar, with medium size, green in colour (Figure 4). Generally the colour of the leaf sheath is green with pink tinge and dark green spines. The mean cane and sugar yields of the cultivar Co 6806 when harvested at the age of 12 months is about 120 ton cane/hectare and 14 ton sugar/hectare, respectively.

The application dose of urea practised under Kenana conditions and also used in the present study was 200 kg of urea/feddan (92 kg nitrogen). Phosphorus at a rate of 50 Kg TSP was added identically to all treatments. Potassium was not applied to any due to high soil K content.

During the process of fertigation technique the irrigation water from the gated pipe was applied to the field from the riser (Field Outlet Pipe) and the dissolved urea was supplied at the site and controlled with a valve

Table 2: (a) Physical properties of Kenana soils; (b) Chemical properties of Kenana soils

(a)

Soil depth (cm)	Mechanical analysis			Field capacity	Saturation	Permanent wilting point	Bulk density
	Sand	Silt	Clay				(gram/cm3)
	%						
0 - 20	11	24	65	46	70	20	1.16
20 - 40	10	22	68	44	66	18	1.25
40 - 60	09	22	69	41	67	21	1.23
60 - 80	09	20	71	42	69	19	1.29

(b)

Soil depth	Soil PH	E.Ce	CaCO3	N	O.C.	Soluble cations		
	Soil:water					Na	Ca	Mg
(cm)	1:5	Dsm−1	(%)			Milli-equivalent/liter		
0 - 20	7.6	0.45	0.5	0.051	0.82	0.2	1.9	0.2
20 - 40	7.8	0.35	1.2	0.055	0.61	0.7	1.6	0.9
40 - 60	8.1	0.33	2.2	0.070	0.50	1.2	1.4	0.6
60 - 80	8.2	0.50	1.9	0.024	0.47	2.0	1.1	0.3

Figure 4: General features of cultivar Co 6806.

connected to the fertilizer tank using water pump and fertilizer agitator. The timing and the amount of water applied and the dissolved urea together with the inflow rates were calculated and recorded every minute for each treatment as presented in Table 3(a) and Table 3(b) (Figure 2 andFigure 3). Storage tank of 4000 litter capacity (with agitator and water pump) for dissolving and injection of the fertilizer into riser of the gated pipes were designed and calibrated to control the dose of fertilizer input per each treatment and the irrigation water supplied to furrows. The fertilizers were injected into the irrigation water by means of pressure differential tank system which is based on the principle of a pressure differential being created by a valve and flow regulator forcing and injecting a varying amount of dissolved fertilizer into the irrigation water (Figure 2 andFigure 3).

Weed control in the present trials was achieved mainly through the application of a mixture of one kilogram Gesapax (Ametryne) and one kilogram Gesaprim (Atrazine) just before the second irrigation and another follow-up treatment as band application of the same chemicals and dose after hilling-up (3.5 - 4 months from planting). Between the two herbicide applications, weeding before hillingup operation was achieved by a dual purpose implement (weeder-reshaper) which control weeds and reshapes the ridges.

Satisfactory and economical control ofTermite damage was achieved by the use of chlorinated hydrocarbon insecticides. This is because of the long lasting residual protection afforded by these insecticides to the sets. "Dursban 48EC" insecticide at the dosage rate of 1.0 litter product per feddan was also applied for the control of termites. Various cultural practices now in use at Kenana have appreciably decreased termites inflicted on sugarcane.

Experiment Procedures and Measurements

In the present study the reference evapotranspiration (ETo) for Kenana estate was computed (Table 1(a)) using the FAO-Penman-Monteith method and the CROPWAT software. CROPWAT was for calculating the reference evapotranspiration (ETc in mm·day^{-1}) for each month of the growing season upon entry of the required meteorological data (Monthly mean minimum and maximum temperature, relative humidity, wind speed, sunshine hours and solar radiation). Sugarcane crop coefficients (kc) for different growth stages were determined following the procedure described by[16] .Then the amounts of water required for the plant was determined as a function of the local climatic data (Table 1(b)).

Soil samples were taken before starting the experiment and just after harvesting the sugarcane crop from the soil depths of 0 - 15; 15 - 30; and 30 - 60 cm at furrow top, middle and the bottom third. The collected soil samples were then analyzed for nitrogen concentration after extracted with water using 10 gram of soil and 10 mL of water and the results were recorded in Table4 During the growing season plant leaf samples were taken from each treatment at the age of six month for the determination of nitrogen concentration (Table 5). The cane was harvested at the age of 12 months and the following growth and yield parameters were determined:

(i) The stem height, (cm) from the soil surface to the top most visible dewlap was recorded in a random sample of 10 plants following the procedure described by [17] .

(ii) The stem diameter, (cm) was measured at harvest time using vernier calliper at 30 cm above the soil surface.

(iii) Cane yield and yield components were determined after harvesting the crop and the data obtained were presented in Table 7(a) and Table 7(b).

Finally, the data obtained from the present study were subjected to the analysis of variance using[18] . The

Table 3: (a) Rates of inflow of irrigation water and dissolved urea used in the process of fertigation technique (first season 2009/2010); (b) Rates of outflow of irrigation water and dissolved urea used in the process of fertigation (second season 2010/2011)

(a)

Treatment	Total irrigation time (hour)	Inflow rate of irrigation water/outlet (L/s)	Injection Time (hour)	Total inflow rate of the dissolved urea/20 outlets (L/s)
Continuous injection of the dissolved urea	6	2.6	6	7.4
Injection of the urea during the first half of irrigation time	6	2.6	3	14.8

Injection of the urea during the second half of irrigation time.	6	2.6	3	14.8
Mechanical application of urea	6	2.6	6	-
Mean	6	2.6	4.5	-

(b)

Treatments	Total irrigation time (hour)	Flow rate of irrigation water/outlet (L/s)	Injection Time (hour)	Flow rate of the dissolved urea (L/s)
Continuous injection of the dissolved urea	8	2.2	8	0.25
Injection of the urea during the first half of irrigation time	8	2.2	4	0.25
Injection of the urea during the second half of irrigation time.	8	2.2	4	0.25
Mechanical application of urea	8	2.2	-	4 bags/ acre
Mean	8	2.2	6	-

Table 4: Average values of the soil nitrogen content for the different treatments

Treatments	Number of composite samples	Soil depth	Soil nitrogen content (%)	
			Crop season 2009/2010	Crop season 2009/2010
Continuous injection of the dissolved urea	18	0 - 60 cm	0.43	0.43
Injection of the urea during the first half of irrigation time	18	0 - 60 cm	0.27	0.32
Injection of the urea during the second half of irrigation time.	18	0 - 60 cm	0.18	0.20
Mechanical application of urea	18	0 - 60 cm	0.14	0.11
Mean			0.25	0.26
CV %			13	15
SE+			0.18	0.04

Table 5: Average values of the leaf nitrogen content for the different treatments

Treatments	Number of samples	Cane age (month)	Leaf nitrogen content (%)	
			Crop season 2009/2010	Crop season 2009/2010
Continuous injection of the dissolved urea	9	6	2.8	2.7

Injection of the urea during the first half of irrigation time	9	6	2.4	2.4
Injection of the urea during the second half of irrigation time.	9	6	2.2	2.0
Mechanical application of urea	9	6	1.5	1.7
Mean			2.2	2.2
CV %			14	16.7
SE+			0.16	0.15

Duncan Multiple Range Test (DMRT) was used to compare the means [19] .

RESULTS AND DISCUSSIONS

In the present study the seasonal amount of water requirements for sugarcane crop were determined as a function of the local climatic data (Table 1(b)). The average values of the irrigation water required to satisfy the sugarcane needs based on the available information on evapotranspiration of the crop were calculated. The accurate determinations of the irrigation water required per each treatment are helpful in managing the dose of fertilizer input per each treatment and the irrigation water supplied to furrows. Accordingly, the injection rates of the dissolved urea in relation to the rate of flow of irrigation water will be defined and monitored by an injector valve. Because when fertilizing through fertigation with a well planned and monitored irrigation schedule, there is less chance of leachable nutrients not being absorbed by the plant. Then the amounts of water required for the plant.

Soil Nitrogen Distribution

The results of the average values of nitrogen concentrations of the soil samples collected and analyzed before starting the fertigation experiment indicated that the average initial soil nitrogen concentrations is about 0.06% with a range of 0.04 to 0.08%, which is considered as very low according to the broad rating of nitrogen percent in soils (Table 6(a)) as given by [20] .

Also the results of soil samples taken just after harvesting the sugarcane crop from the soil depths of 0 - 15; 15 - 30; and 30 - 60 cm at furrow top, middle and the bottom third were recorded and presented in Table3 The collected soil samples were then analyzed for nitrogen concentration after extracted with water using 10 gram of soil and 10 mL of water and the results were recorded inTable4 In both seasons the obtained results revealed that, following fertigation, the nitrogen concentration along the length of run was highest in the top 15 cm of soil for most treatments and decrease with depth. The present study also indicated that when the dissolved urea (46% N) was injected during the entire irrigation time (100% treatment), soil nitrogen concentrations for each soil depth in the furrow (top, middle and bottom third) were similar along the entire length of the field. Whereas, when the dissolved urea injected during the first half of the irrigation, nitrogen concentrations in the soil were higher at the lower end of the field than at the upper end. Greater total nitrogen at the lower end than at the upper end occurred because of the longer infiltration opportunity time of nitrogen-amended water at the lower end. Soil nitrogen concentrations along the furrows where nitrogen injection was made during the last half of the irrigation were higher at the head end of the field, with almost no nitrogen at the lower end of the field. This is due to the early rapid infiltration of un-amended water in the furrows at the upper end of the field. Injection during the last half of the irrigation resulted in little total nitrogen at the tail end of the field and the largest accumulation of total nitrogen at the head end of the field. [21] using similar treatments for level furrows, found similar results and concluded that injection of nitrogen concentrations during the entire irrigation time of level

Table 6: (a) Explains the broad rating of nitrogen percent in soils as given by (Landon, 2014)*; (b) The rating of sugarcane leaf total nitrogen percent as given by the Agricultural Extension of Florida, USA.

(a)

Nitrogen content (%), Kjeldahl Method	Rating
>1.0	very high
0.5 - 1.0	high
0.2 - 0.5	medium
0.1 - 0.2	low
<0.1	very low

(b)

Country	Reference	Optimum leaf nitrogen percent
USA	(McCray & Mylavarapu, 2010)	2.0 - 2.6

Table 7: (a) Yield and yield components obtained during crop season 2009/2010 (first season); (b) Yield and yield components obtained during crop season 2010/2011 (second season)

(a)

Treatment	Plant height (cm)	Plant diameter (cm)	Cane yield (TCF)
Continuous injection of the dissolved urea	298	2.5	45.0
Injection of the urea during the first half of irrigation time	282	2.3	43.4

Injection of the urea during the second half of irrigation time.	275	2.0	42.0
Mechanical application of urea	255	1.9	36.4
Mean	278	1.6	40.9
CV%	2.36	10.23	6.67
SE+	2.67	0.13	1.12

(b)

Treatment	Plant height (cm)	Plant diameter (cm)	Cane yield (TCF)
Continuous injection of the dissolved urea	305	2.6	41.0
Injection of the urea during the first half of irrigation time	292	2.1	38.4
Injection of the urea during the second half of irrigation time.	282	1.5	34.0
Mechanical application of urea;	275	1.5	32.9
Mean	289	1.9	36.6
CV%	13	19.3	6.74
SE+	4.75	0.16	1.06

furrows also produced the highest uniformity for nitrogen concentrations, but that injection during the first half of the irrigation time was the next best option. [22] also found that injection during last half of the irrigation produced significantly lower uniformity than injection during the entire irrigation or first half.

With respect to the mechanical application of urea using fertilizer distributor and rigid Tyne cultivator (conventional method), fertiliser was first broadcasted and then incorporated in the furrows. Due to root pruning, fertiliser incorporated in the furrows, away from the sides of the ridges, would not be easily accessible to the roots of the ratoon cane. Additionally, it was observed that the rigid Tyne cultivator did not completely cover the broadcasted fertilisers (Figure 4) and that would eventually lead to loss of urea through volatilization. Moreover, it was also observed that during broadcasting, fertiliser was sometimes drifted to the top of cane rows, where they could not be incorporated and thus provided an additional source for losing urea to the atmosphere. Accordingly, very poor nitrogen use efficiency was obtained that resulted in lower yield and yield components (Table 7(a) and Table 7(b)). [23] revealed that adoption of proper method of fertiliser application is essential to minimize nutrient losses from the soil and for increasing fertiliser use efficiency. According to [24] fertilisers should be placed close to the root zone to enable the roots to derive its benefit immediately. Beside increasing cane yield, proper placement also reduces volatilization losses of nitrogenous fertilisers. [25] mentioned that when applied to the soil, urea $[CO(NH_2)_2]$ is converted to ammonia (NH_3) by an enzyme called urease, then to ammonium (NH_4). If that reaction occurs at the soil surface, some of the nitrogen will be lost to the atmosphere. However, if it occurs in soil the ammonia will convert to ammonium and will be held by the clay particles.

Leaf Nitrogen Distribution

Results of the present study indicated that improvement in leaf nitrogen due to continuous injection of urea were better in both seasons and significantly ($P < 0.05$) increase yield and yield components due to that continuous injection of nitrogen fertilizer satisfy the nitrogen demand of sugarcane crop leading to better uptake of the dissolved nitrogen fertilizer than the other treatments (Table 5), whereas the mechanical application approach leading to the worst nitrogen uptake.

In both seasons the results of plant leaf analysis data were then compared with the international references mainly, the rating of sugarcane leaf total nitrogen percent as given by the Agricultural Extension of Florida, USA (Table 6(b)).

Stem Height and Diameter (cm)

Data on the effect of different fertigation strategies on stem height are given in Table 6(a) andTable 6(b). In both seasons the results show that, continuous injection of urea significantly ($P < 0.05$) increase stalk height as compared to the other diameter (Table 7(a) and Table 7(b)) showed that the different treatments had no significant effect on stem diameter in both seasons. The insignificant reduction in stem diameter associated with mechanical application of urea (in both seasons) may be due to the fact that the crop was grown under deficit nitrogen conditions that restricted stem thickness. These parameters were believed to be the first to be affected by nitrogen stress on treatments. However, mechanical urea application produced the poorest stalk height, because the crop nitrogen requirement was not satisfied. The effect of different fertigation strategies on stem the other hand, the highest values of stem diameter occurred in plants fertigated at 100% irrigation time.

Cane Yield

Cane yield and yield components were determined after harvesting the crop and the data obtained were presented in Table 7(a) and Table 7(b). Results obtained indicated that in both seasons continuous injection of the dissolved urea (46% N) during the entire irrigation time (100%), significantly ($P < 0.05$) increased the average cane yield (ton/feddan), whereas the lowest cane yield was recorded when urea was applied mechanically. The increase in cane yield and yield components with the continuous injection of the dissolved urea as compared to the other treatments could be because of a more uniform water distribution during the whole irrigation time resulting in an improved distribution of the fertilizer.

CONCLUSIONS

Following the evidence presented in this paper, the following conclusions are proposed:

When the dissolved urea was injected during the entire irrigation time, soil nitrogen concentrations for each soil depth in the furrow (top, middle and bottom third) were similar along the entire length of the field. Nevertheless, when the dissolved urea was injected during the first half of the irrigation, nitrogen concentrations in the soil were higher at the lower end of the field than at the upper end because of the longer infiltration opportunity time of nitrogen-amended water at the lower end. Soil nitrogen concentrations along the furrows where nitrogen injection was made during the last half of the irrigation were higher at the head end of the field, with almost no nitrogen at the lower end of the field. This is due to the early rapid infiltration of un-amended water in the furrows at the upper end of the field. Injection during the last half of the irrigation resulted in little total nitrogen at the tail end of the field and the largest accumulation of total nitrogen at the head end of the field.

Also it is concluded that continuous injection of the dissolved urea during the entire irrigation time produced the best distribution uniformity of added nitrogen and increase in the plant height, diameter and cane yield, whereas injection of the dissolved urea during the first half of the irrigation time was the next. However, these differences were only significant when comparing the mechanical application of urea with fertigation treatments. It worth emphasizing that the differences were not significant between the three fertigation treatments. Additional field studies are needed with different operational conditions to develop more complete guidelines for fertigation using gated-pipe furrow irrigation system.

RECOMMENDATIONS

Based on two season's results the author recommends that, regulation of the injection of fertilizer to match the flow of irrigation water is so precise that application rates are substantially lowered. Accordingly, fertilization of the sugarcane crop (ratoons) should be carried out via fertigation strategy mainly through injection of the dissolved urea (46% N) during the entire irrigation time (100% treatment). Because this treatment produced the best distribution uniformity of added nitrogen compared to the injection during the first or the second half of the

Table 8: Benefits gained and cost saved due to introduction of fertigation

Parameter	Fertigation technique	Conventional fertilizer application method
Uniformity of fertilizer application	Distribute the dissolved urea more evenly throughout the wetted root zone, with less labour	Intensive, time consuming and do not provide proper fertilizer placement on ratoon fields.

Area fertilized per day	Depend on the area to be irrigated/day	Using fertilizer applicator, only 100 feddan/10 hours a day.
Tractor horsepower required for application of the fertilizer.	Lowest fuel consumption and energy were required only to transport the fertilizer storage tank once. Fertigation leads to saving of tractor horse-power and operation cost and time.	Broadcasting of fertilizers in separate operations represented an inefficient use of tractor power and time. The broadcasters required only partial power to operate the metering mechanism (0.05 kW). The weight of filled fertiliser hoppers added 1.68 kW to the rolling resistance of the tractor. Maximum power for a broadcaster was therefore around 1.73 kW. This almost means that the power of the tractor was only utilized to transport itself across the field.
Incorporation and covering of the applied fertilizer	This operation is not required accordingly, more energy and power were saved.	The average output/10 hours a day when using the rigid Tyne cultivator is about 30 feddan. This operation required about 12.32 (kW-h)/ha

irrigation time. Also fertigation leads to elimination and saving of tractor horse-power and operation cost and time (Table 8).

REFERENCES

1. Moberly, P. and Meyer, J.H. (1984) Soils: A Management Factor in Sugarcane Production in the South African Sugar Industry. Proceedings of the South African Sugar Technologist's Association, 58, 192-197.

2. Bakker, M. (1999) Sugarcane Cultivation and Management. Kluwer Academic/Plenum Publishers, New York. http://dx.doi.org/10.1007/978-1-4615-4725-9
3. Meyer, J.H. and Wood, A.W. (2000) Soil Management Research for Sustainable Cane Production in the 21st Century. In: Wilson, J.R., Hogarth, D.M., Campbell, J.A. and Garside, A.L., Eds., Sugarcane: Research towards Efficient and Sustainable Production. CSIRO, Brisbane, 343-358.
4. Papadopoulos, I. (1996) Micro-Irrigation Systems and Fertigation. In: Pereira, L.S., et al., Eds., NATO ARW on Sustainability of Irrigated Agriculture, Kluwer Academia Publishers, New York, 309-322.
5. Brady, N. and Weil, R. (1999) The Nature and Properties of Soils. Prentice Hall, Upper Saddle River.
6. Plaster, E. J. (2003) Soil Science and Management. 4th Edition, Delmar Learning, USA.
7. Burt, C., O'Connor, K. and Ruehr, T. (1995) Fertigation. California Polytechnic State University, Irrigation Training and Research Center, San Luis Obispo.
8. Beth, F. and Filters, A. (1981) Fertigation Gains Increasing Acceptance. Irrigation Farmer, 8, 2.
9. Howell, T.A., Stevenson, D.S., Aljibury, F.K., Gitlin, H.M., Wu, I.P., Warrick, A.W. and Raats, P.A.C. (1980) Design and Operation of Trickle (Drip) Systems. In: Jensen, M. E., Ed., Design and Operation of Farm Irrigation Systems, ASAE, St. Joseph, 663-717.
10. Playán, E. and Faci, J.M. (1997) Border Fertigation: Field Experiments and a Simple Model. Irrigation Science, 17, 163-171. http://dx.doi.org/10.1007/s002710050035
11. Boldt, A.L., Watts, D.G., Eisenhauer, D.E. and Schepers, J.S. (1994) Simulation of Water-Applied Nitrogen Distribution under Surge Irrigation. Transactions of the ASABE, 37, 1157-1165. http://dx.doi.org/10.13031/2013.28189
12. Garside, A.L. and Nable, R.O. (1996) Sugarcane Growth and Yield Comparisons in Paired Old and New Land Sites. In: Wilson, J.R., Hogarth, D.M., Campbell, J. and Garside, A.L., Eds., Sugarcane: Research towards Efficient and Sustainable Production, CSIRO Division of Tropical Crops and Pastures, Brisbane, 248-250.

13. Burt, C.M. (1995) Irrigation Evaluations. In: The Surface Irrigation Manual, Waterman Industries, Inc., Exeter, 14.1-14.21.
14. Bouwer, H. (1990) Agricultural Chemicals and Groundwater Quality. Journal of Soil and Water Conservation, 45, 184-189.
15. Watts, D.G., Ostermeier, K.A., Eisenhauer, D.E. and Schepers, J.S. (1993) Fertigation by Surge Irrigation on Blocked-End Furrows. Proceedings of Conference on Agricultural Research to Protect Water Quality, Minneapolis, 21-24 February 1993, 523-526.
16. Farbrother, H.G. (1971) Crop Water Requirements (CWR) of Sugarcane and Net Irrigation Requirement in the North West Sennar Area. Agronomy and Crop Physiology Section. Gezira Research Station.
17. Clements, H.F. (1980) Sugarcane Crop Logging and Crop Control: Principles and Practices. University Press of Hawaii, Honolulu.
18. SAS Institute (1990) SAS/STAT User's Guide 1990 Edition. SAS Institute, Inc., Cary.
19. Gomez, K.A. and Gomez, A.A. (1984) Statistical Procedures for Agricultural Research. 2nd Edition, John Wiley and Sons Inc., New York.
20. Landon, J.R. (2014) Booker Tropical Soil Manual: A Handbook for Soil Survey and Agricultural Land Evaluation in the Tropics and Subtropics. In: 2000 Symposium: Sugarcane: Research towards Efficient and Sustainable Production, Routledge, London, 237-240.
21. Abbasi, F., Simunek, J., van Genuchten, M.T., Feyen, J., Adamsen, F.J., Hunsaker, D.J., Strelkoff, T.S. and Shouse, P. (2003) Overland Water Flow and Solute Transport: Model Development and Field-Data Analysis. Journal of Irrigation and Drainage Engineering, 129, 71-81. http://dx.doi.org/10.1061/(ASCE)0733-9437(2003)129:2(71)
22. Adamsen, F.J. and Rice, R.C. (1995) Nitrate and Water Transport as Affected by Fertilizer and Irrigation Management. Proceedings of 21st Century Conference: Clean Water—Clean Environment: Vol. II. Nutrients, 1-4.
23. Watts, D.G. and Schepers, J.S. (1995) Fertigation to Reduce Nitrate Contamination of Groundwater. Proceedings of 21st Century Conference: Clean Water—Clean Environment: Vol. II. Nutrients, 239-242.

24. Verma, R.S. (2002) Sugarcane Ratoon Management. International Distribution Company, India.
25. Deer and Company (1993) Soil Management. Deer and Company Service Publications, Moline.

Chapter 2

Impact of Changes in Diet on the Availability of Land, Energy Demand, and Greenhouse Gas Emissions of Agriculture

Karin Fazeni and Horst Steinmüller

Energy Institute at the Johannes Kepler University (JKU Linz), Altenbergerstrasse, 69, Linz, 4040, Austria

ABSTRACT

Background

Recent scientific investigations have revealed a correlation between nutrition habits and the environmental impacts of agriculture. So, it is obviously worthwhile to study what effects a change in diet has on land use patterns, energy demand, and greenhouse gas emissions of agricultural production. This study calculates the amount of energy and emission savings as well as changes in land use that would result from different scenarios underlying a change in diet.

Methods

Based on the healthy eating recommendations of the German Nutrition Society, meat consumption in Austria should decrease by about 60%, and consumption of fruits and vegetables has to increase strongly.

Results

This investigation showed that compliance with healthy eating guidelines leads to lower energy demand and a decrease in greenhouse gas emissions, largely due to a decrease in livestock numbers. Furthermore, arable land and grassland no longer needed for animal feed production becomes redundant and can possibly be used for the production of raw materials for renewable energy. The scenario examination shows that in the self-sufficiency scenario and in the import/export scenario, up to 443,100 ha and about 208,800 ha, respectively, of arable land and grassland are released for non-food uses. The cumulative energy demand of agriculture is lower by up to 38%, and the greenhouse gas emissions from agriculture decrease by up to 37% in these scenarios as against the reference situation.

Conclusion

The land use patterns for the scenario demonstrate that animal feed production still takes up the largest share of agricultural land even though the extent of animal husbandry decreased considerably in the scenarios.

INTRODUCTION

Agriculture has various impacts on the environment. One of the most obvious impacts is the emission of methane [CH_4], nitrous oxide [N_2O], and other greenhouse gases from ruminant animals and manure management, the application of mineral and organic fertilizers [1], and soil management practices [2,3]. These greenhouse gas emissions contribute significantly to climate change in line with their global warming potential [1]. In addition, agriculture also contributes to emissions by the consumption of energy, both directly, in the operation and maintenance of plant and machinery used to cultivate cropland and maintain livestock housing, and indirectly, in the form of manufactured mineral fertilizers and pesticides. The level of energy consumption and greenhouse gas emissions depends on the production system, for example, whether organic or not, and on the product mix, i.e., the mix of crops and livestock. It has been shown that organic farming consumes less energy and contributes less to greenhouse gas emissions than conventional agriculture because of the abandonment of fossil-fuel-derived nitrogen and synthetic pesticides [4-11]. Besides the approach to input use, soil management practices, such as tillage, irrigation, use of cover crops [2] in cropping systems, and storage of slurries and manures in livestock systems, also influence greenhouse gas emissions from agriculture. In the context of choice of the cropping system, crop rotation has a strong influence on emissions. For example, adapting crop rotations to include more perennial crops, thereby avoiding use of bare and fallow land, reduces greenhouse gas emissions from agriculture by accumulating soil carbon stocks [3]. Animal husbandry is recognized to have higher energy consumption and therefore has more greenhouse gas emissions than arable agriculture. In fact, 18% of the global greenhouse gas emissions stems from livestock production, whereby CH_4 from enteric fermentation in ruminant animals is a major contributor, followed by N_2O and carbon dioxide [CO_2][12]. The high levels of animal protein found in modern western diets does not only affect land use[a], but is also a significant driver of current levels of energy consumption and greenhouse gas emissions of agriculture [4,5,7-13]. The correlation between nutritional habits and emissions from agriculture has already been shown in other studies with different geographical foci [14, 15].

[a]Production of 1 kg of beef requires an area of 20.9 m^2, while 1 kg of cereals only requires about 1.4 m^2 of arable land [54].

The high land requirements of livestock production, coupled with a growing demand for meat in developing countries, raise the specter of shortages of arable land over the next few decades [16]. Indeed, some authors have also questioned whether it will be at all possible to feed so many animals in the future [17]. In addition, there is a growing demand for land for the production of renewable energy feedstocks [18]. As the markets for crop feedstocks for bioenergy and biofuels grow [19], arable land is bound to be reallocated to meet these new demands [19]. Demand for feedstock for bioenergy can affect food supplies in two ways: first, by diverting land to the production of non-food crops and second, by diverting food and feed crops to renewable energy uses. Both of these outcomes constrain food and feed supply, and this in turn impacts on prices [20]. The years 2007 and 2008 witnessed very significant food price rises, which especially affected the developing countries. One of the major factors for these price increases was the demand for maize for bioethanol production. Although demand for biofuel feedstocks is only one factor pushing food prices up, alongside droughts and bad harvests, biofuel production exacerbated the situation [21]. Among experts, there is an agreement that biofuels have an important role in reducing greenhouse gas emissions, and with energy prices rising and public policies supporting their use, the demand for biofuels will continue to grow. The challenge for governments is to find approaches that can accommodate the competing demands of the food and biofuel sectors. One possible future option is to make biofuels from a cellulosic feedstock which does not compete with food production [22]. Another approach is to encourage a shift to a diet with less meat intake [23]. Stehfest et al. [12] showed that land which becomes redundant because of changed nutritional habits could possibly be used for energy crop production. Table 1 gives estimates of the area which currently might be used for renewable energy feedstock production in Austria, together with a number of scenarios of land use change as modeled in this study.

Table 1: Area available for renewable energy feedstock production in Austria currently

Arable land and grassland available for renewable energy feedstock production in Austria				
Baseline situation in 2006	Estimated potential in 2020 for a national Biomass Action Plan	Estimated potential in a Biomass Resource Potential Study in 2020	Estimated potential in self-sufficiency scenario (maximum)	Estimated potential in import/export scenario (maximum)
55,000 ha	1,011,000 ha	455,000 ha	443,100 ha	208,800 ha

The said available area for renewable energy feedstock production is also under a number of scenarios of land use exchange. The data come from BRAINBOWS [53] and from the authors' own calculation.

Fazeni and Steinmüller *Energy, Sustainability and Society* 2011 1:6, doi: 10.1186/2192-0567-1-6

Both the correlation between the choice of diet, agricultural greenhouse gas emissions, and energy consumption and the land use competition between food and energy crops have already been discussed in past publications, e.g., [12,17,24,25]. A similar work by Freyer and Weik [13] has been done for Austria. They found out that the CO_2e emissions related to a nutritional recommendation by the German Nutrition Society [DGE] are about 1,031 kg per capita and year.

Although a good deal of research has been done on these topics, only a few studies, e.g., [12], have investigated the impacts of a change in diet on agricultural greenhouse gas emissions, energy consumption, and land use in an integrated way for a whole country. The present study addresses this deficit by analyzing the impacts of a change in diet on land use, energy consumption, and the emissions of Austrian agriculture, together with the potential for producing renewable energy feedstocks using redundant land. A major aim of this work is to show the complex interactions between food demand, agriculture, emissions, and renewable energy production.

Finally, we estimate how much renewable energy feedstocks may be produced in Austria without competing with food production in the case of changed nutritional habits. This approach also makes it possible to discuss whether changed nutritional habits are an available future option to limit the extent of competition between food production and renewable energy feedstock production. The results of this work may provide starting points for an integrated policy addressing the diet of the population, agriculture, and renewable energy production.

MATERIALS AND METHODS

The life cycle assessment [LCA] (EN ISO 14040:2006) approach was chosen to quantify the cumulative energy demand [CED] of and the related greenhouse gas emissions from the conventional agriculture in Austria. The LCA method seems to be appropriate for reaching this goal because the CED and the corresponding emissions are an integrated component of every LCA study [26].

There is no agreed standard for calculating energy balances in the context of agriculture, with various approaches documented in the literature. In terms of analyzing the energetic aspects of agro-ecosystems, a hierarchy of methods exists. The approach adopted for this study is a mechanistic, technical one, where all energy inputs are traced into an agricultural system as physical material flows [27]. The involvement of material flows shows again that the application of the EN ISO 14040:2006 method for this work is appropriate. As a method for measuring the energy demand of agriculture, CED was chosen. The CED was developed in the 1980s and has played an important part in impact assessment since the early development of LCA. Because CED aggregates all forms of energy consumed over the whole life cycle including losses, it is a sum parameter, i.e., a meaningful parameter used to quantify the primary energy demand of a system and its upstream stages. CED is derived from inventory analysis, where mass and material flows have to be known [28], so it does not depend on any assumptions and their associated uncertainties made in impact assessment [29]. CED is also an appropriate yardstick for comparing products [30] and scenarios [31, 32]. According to EN ISO 14040:2006, LCA is divided into four steps: goal and scope definition, inventory analysis, impact assessment, and finally, interpretation. The approach taken in this study stops just short of a full conventional LCA, but nevertheless, it consists of a life cycle inventory analysis survey although an impact assessment is carried out for the impact categories, global warming potential and CED. The impact assessment steps of characterizing and classifying inventory results (EN ISO 14040:2006) are necessary to show the results in CO_2 equivalent and the CED [33].

Employing the LCA method on the entire Austrian agricultural system posed some difficulties because LCA methods developed for agriculture are mostly designed for use at farm level [34]. Other agricultural LCA approaches are tailored to just a single agricultural sector [35] or a single agricultural product [36, 37]. Therefore, a manageable approach had to be developed to employ the LCA method on the whole of Austrian agriculture. As a result, to reduce complexity, Austrian agriculture is treated as a single average farm. This average farm cultivates all Austrian farmland, grows all demanded crops, and breeds all demanded animals. Crop rotation is determined by the current pattern of crop cultivation, both in the reference case and in the scenario analysis. As a consequence, the LCA can be thought of as being performed at the 'notional' farm level.

Methodology of Energy Accounting

Definition of the Goal and Scope for Energy Accounting in Conventional Agriculture in Austria

In line with the goal definition and principles of LCA (ISO, 2006) and following the approach taken by Hülsbergen et al. [38], the agricultural production process chain, i.e., all relevant upstream stages of agricultural production (such as the production of fertilizers and pesticides and the upstream stages of energy supply), is taken into account for current energy accounting. On the downstream side, the farm gate is treated as the system boundary. So, transporting crops from the field to the farmyard takes place within the system, but not transporting or processing beyond that point. This ensures the same system boundary for animal husbandry and crop production. The construction and maintenance of agricultural infrastructure such as farm buildings and machines are not within the system boundaries. Other inputs not taken into account are solar energy used by growing crops and energy inputs to human labor.

Figure 1 is a simplified diagram of the LCA system boundaries. The picture shows the main inputs into the Austrian agricultural production system, consisting of mineral fertilizers, organic fertilizers, pesticides, electricity, diesel fuel, thermal energy, and animal feed from industry. The stages of processing the agricultural operating resources are taken into account in the calculations. The CED of seeds is estimated as the CED used for the part of current crop production that is retained for use as seeds in the next cultivation period. In Austrian agriculture, seed retention ranges from 0.5% to 7% depending on the crop. A transport process between field and farm takes place. Cultivated crops and grass forages are brought from the field to the farm, where they are either exported off the farm or fed to livestock. The animal products accounted for are meat, milk, and eggs. The processing stages of food transport off the farm processing are not taken into account.

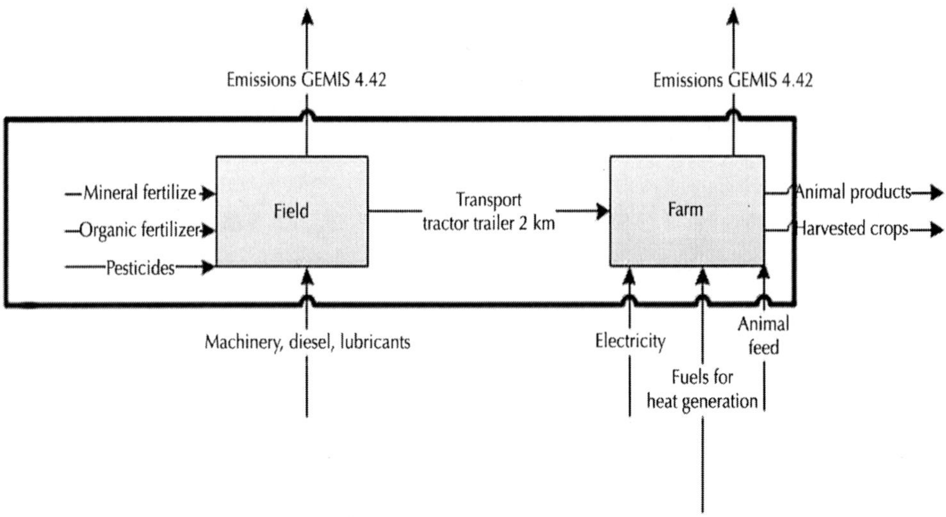

Figure 1: LCA system boundaries. The data are based on the authors' calculation.

Life Cycle Inventory Analysis for Austrian Agriculture

A life cycle inventory analysis characterizes the juxtaposition of the quantified inputs and outputs [39] of agricultural production. In the present case, the inputs are fertilizer, pesticides, animal feed, and energy; the outputs are the emissions involved in consuming these factors of production. The software model Global Emission Model for Integrated Systems [GEMIS] (Version GEMIS Austria 4.42-2007, Institut für angewandte Ökologie e.V., Vienna, Austria) [40] was used to quantify the associated emissions and CED.

GEMIS comprises a lot of different agricultural processes including the correlation of energy demands and CO_2e emissions, describing both plant production and animal production. Consequently, GEMIS makes it possible to take all relevant agricultural processes into account, including energy demand and the associated emissions from upstream stages such as mineral fertilizer and synthetic pesticide production. Not all processes relevant to calculating the CED of Austrian agriculture were available in GEMIS for carrying out process chain analysis; so, some processes had to be modeled, and other processes had to be adapted to Austrian agricultural conditions. For adapting the processes in GEMIS, special data on fertilizer and pesticide application as well as data on the direct energy demand of Austrian agriculture had to be obtained. Data on fertilizer and pesticide application were provided by the Austrian Association for Agricultural Research. Details of the data set used and methods of data generation are described in the literature [11]. For determining the average rates of fertilizer and pesticide application in Austrian agriculture, guidelines published by the Austrian Ministry of Agriculture were used. Other data, especially concerning the direct energy consumption of agriculture, were obtained from the literature [41-46] and from stakeholder interviews. For more details on this procedure and the data that were derived, read about the study of Zessner et al. [47]. In GEMIS, a separate process exists for each agricultural product. As a first step, the CED and emissions are calculated for each agricultural product separately. As GEMIS outputs are denominated per ton of a specific product, the outcome has to be multiplied by the whole production volume determined for the baseline situation and for the scenarios. By this means, the CED and CO_2e for the whole Austrian production of a specific crop or animal product are calculated. Aggregating these results yields the entire CED and greenhouse gas emissions for the whole of Austrian agriculture.

Scenario Definition and Description

Scenario Definition: Common Assumptions

Initially, it has to be clarified that the scenarios examined in this paper are retrospective. By this means, uncertainties concerning future states of drivers of change such as increasing technical efficiency, demographic changes in Austria, or developments in agricultural policy are avoided. These influencing parameters stay constant *vis-à-vis* the baseline period, i.e., the average of 2001 to 2006. As already stated, in all the scenarios the impacts on the existing conventional agricultural system of changing nutritional habits among the population of Austria are examined. The scenarios have been developed on the assumption that only conventional farming methods are used [47].

For the purposes of scenario analysis (all scenarios), it is assumed that dietary change involves the compliance of the Austrian population with the recommendations of the DGE. Today, meat consumption in Austria exceeds the levels recommended in healthy eating guidelines. According to the DGE recommendations, meat consumption of the average Austrian inhabitant would need to decrease by about 60% of today's level of 57 kg per capita per year. This will result in a shift to more plant-based nutrition, with the consumption of fruits and vegetables increasing by about 50% and 60%, respectively (for a more detailed information, read more on the study of Zessner et al. [48]).

The DGE recommendations refer to specific product groups such as fruits. To calculate the amount of food needed for the population of Austria in one year, the average recommended daily or weekly intake of a specific food product was taken. Next, the amounts of agricultural products, such as milk, eggs, cereals, and oil, needed to meet the demand for healthy nutrition were determined. To calculate total agricultural production, net food consumption was derived using correction factors for each food category. Net food consumption determines how much livestock and arable land is needed to produce all the agricultural goods in demand. Animal feed amounts were derived from the specific animal feed demand per animal category. A distinction was made between ruminant animals and monogastric animals. This calculation yielded the area of arable land and grassland needed for animal feed production [47].

The starting point of each scenario is a change in diet among the population of Austria in line with the DGE recommendations. This change in diet between the baseline situation and the scenarios is presented in Table 2.

Table 2: Consumption of food by product categories in the baseline situation and the scenarios

	Baseline situation	Scenario situation
Product categories	[kg/per capita/annum]	
Meat	56.8	23.4
Eggs	11.8	9.5
Milk and milk productsa	257.0	279.9

Fish	9.8	0.4
Cereals/rice/potatoes	114.6	129.7
Fruits	58.6	91.3
Vegetables	89.6	146.0
Vegetable oils	9.7	6.8
Sugar	33.0	18.3

[a]Raw milk equivalent; The data are based on the authors' own calculation which is based on the study of Zessner et al.[48].

Fazeni and Steinmüller *Energy, Sustainability and Society* 2011 1:6, doi: 10.1186/2192-0567-1-6

Agricultural production has to be adjusted to these changes in commodity demand. In the case of meat consumption, it is assumed that consumption of all meats decreases to the same extent. Although common healthy eating guidelines recommend eating more white meat than red meat, this study assumes that the shares of the various sorts of meat stay the same because people would still prefer red meat. The consumption and production of alcoholic beverages are left unchanged because no commonly accepted recommendation is available from nutrition scientists. As the efficiency of agricultural production is assumed to be the same as in the baseline period, the same amount of resources is consumed in producing a given product conventionally as in the baseline situation. Agricultural production is not expanded to forest areas, and the amount of fallow land cannot increase beyond the level observed in the baseline period [47].

In the import/export scenario, net imports change in proportion to the change in food and animal feed demand in Austria. An exception is made in the case of saltwater fish because it is assumed that there is no potential, in view of depleted fish stocks, to increase the supply of fish from the world's oceans. The lack of omega-3 and omega-6 fatty acids is made good with vegetable oils. In this scenario, exports stay at the same level as in the baseline situation in absolute terms. Currently, about 26,000 t of meat and 361,700 t of milk are exported per year, with most of the meat exported being beef [47]. Once the main assumptions for the scenario definition have been settled, the different scenarios and sub-scenarios examined in this work can be described.

The scenario development largely depends on the assumed self-sufficiency in agricultural production. Even in the baseline situation, Austria is already close to self-sufficiency in some agricultural goods. Self-sufficiency in grain in Austria was about 100% and self-sufficiency in potatoes, about 96% in 2005/2006; self-sufficiency in meat in Austria was about 106% and in milk, about 136% in the year 2006. Austria is much further from self-sufficiency in oil seeds (59%), fruits (69%), and vegetables (57%). Where Austria is quite close to self-sufficiency, the simplifying assumption is made that the country is 100% self-sufficient in these products. Where full self-sufficiency in agricultural goods is assumed, some consumption assumptions are also required. For example, because rice plays a role in the diet of the average Austrian and because domestic rice cultivation is not possible, in the scenario, modeling has to be replaced by other starchy foods such as potatoes and cereals. Full self-sufficiency also means that the amount of fish recommended by the DGE cannot be produced in Austria, so the Austrian population is assumed to be supplied with omega-3 and omega-6 fatty acids in the form of linseed oil, walnut oil, and rape seed oil. Again, in the full self-sufficiency scenario, tropical and subtropical fruits are replaced by domestic fruits. The substitution was done in line with the ratio of domestic fruit types actually consumed. For example, as apples have the largest share of fruit consumption in Austria, most tropical and subtropical fruits are replaced by apples [47].

In determining agricultural production, crop rotation constraints have to be taken into account. In this case, the following crop rotation constraints were assumed for conventional agriculture in Austria: the share of grains in crop rotation should be < 65%; the share of oil seeds, < 25%; the share of legumes, < 25%; and the share of root crops, < 50%. These constraints are crucial for determining the energy feedstock crops to be produced in the various scenarios [47].

Using the assumptions outlined above, the following scenarios were developed [47] (see Figure 2):

- *'Self-sufficiency' scenario.* The central assumption in this scenario is that Austria is 100% self-sufficient in agricultural goods. No agricultural products are imported or exported.
- *'Import/export' scenario.* In contrast to the self-sufficiency scenario, agricultural goods are imported and exported in the import/export scenario. Exports stay at the same level as in the

baseline situation from 2001 to 2006. Imports are adapted to the new demand pattern in Austria after the change in diet. These assumptions are scenario constraints, not a market outcome.

- For both the self-sufficiency scenario and the import/export scenario, the following sub-scenarios are examined. In conclusion, six sub-scenarios are calculated.
- *Sub-scenario a*. In this sub-scenario, the agricultural production is limited to food production. The production of renewable energy feedstocks is constant at the level already produced in the baseline situation (2001 to 2006).
- *Sub-scenario b*. In addition to food production, agriculture produces renewable raw materials for supplying itself with bioenergy and biofuels on released arable land and grassland. Furthermore, biofuels for fulfilling the transport fuel renewable obligation as per mandate of the European Parliament [49] are produced.
- *Sub-scenario c*. This sub-scenario assumes maximum energy production from agricultural raw materials based on first generation bioenergy and biofuel technologies. The general assumption is that all the redundant agricultural land is used for energy feedstock production.

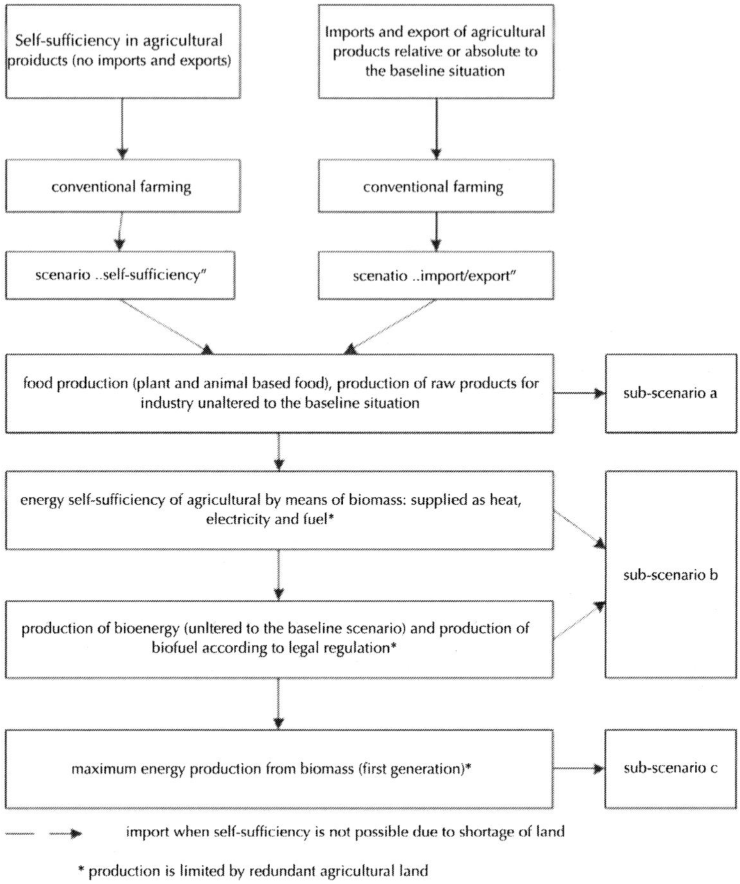

Figure 2: Scenario description. The data are from the authors' calculation which is based on the study of Zessner et al. [47].

Determining the Production of Renewable Energy Feedstocks in the Sub-Scenarios Self-Sufficiency (a, b, and c) and Import/Export (a, b, and c)

One of the main outputs of this analysis is the quantity of renewable energy feedstocks produced under the conditions of the various sub-scenarios. The volumes produced will obviously be dependent on the area of land made available due to decreased meat production. It was assumed that where arable land and grassland are released due to falls in livestock production, this occurs evenly all over Austria. This assumption is necessary because of uncertainties over the likely real world location of the land that was released. It is assumed that this redundant grass is harvested as a feedstock for bioenergy production.

Due to the necessity of crop rotation, oilseed (rape and sunflower) cultivation cannot be expanded in any of the scenarios. The cultivation areas currently observed, 59,000 ha of which is currently used to supply biodiesel feedstocks, are retained as upper constraints. In the scenario analysis, it is assumed that any biodiesel produced is used only within agriculture.

Free grassland and silage maize are used for biogas production. There are two different technical options for the use of biogas for heat and electricity production. One option is combined heat and power generation, and the other option is to feed upgraded biogas into the natural gas grid for power generation in a large-scale gas-power station. A mix of these two technologies is also possible.

In the case of bioethanol production, i.e., to meet the feedstock requirements of the national bioethanol plant, a maize wheat ratio of 1:1 is assumed. As a result, based on average yields, 52,000 ha of wheat and 25,000 ha of maize would be needed to meet the demand.

RESULTS

Because the baseline situation and scenario results that follow are derived from a process chain analysis carried out by means of GEMIS, it is important to show how upstream stages, such as fertilizer production, contribute to a single agricultural production process. To facilitate this, the results are presented by the agricultural sector for each scenario and also for the baseline situation.

The Contribution of Upstream Processing Stages to CO_2e and CED

As mentioned above, CED has been chosen as the most appropriate measure to quantify the energy and emission balance of Austrian agriculture in this study because it includes all primary energy used throughout the life cycle. This measure permits the contribution of upstream processing stages, such as fertilizer production, to CO_2e emissions to be estimated. Rather than try to estimate the emissions of all upstream processing, the upstream contribution to wheat production was chosen as an exemplar for the contribution of upstream production stages in general. Wheat was chosen due to its heavy reliance on mineral fertilizer production, which accounts for a large part of the upstream CO_2e contribution of conventional agricultural production. Accounting for all sources, the production of 1 t of wheat yields a CED of 676 kWh and emissions of 360 kg of CO_2e, where 31% of the CED and 27% of the CO_2e emissions are attributable to the processing stage of mineral fertilizer production. It is therefore safe to say that the CED and CO_2e emissions of agricultural products are closely related to the use of mineral fertilizers. It should be mentioned that the use of mineral fertilizers and pesticides in the scenarios stays at the same level as in the baseline situation.

CED and CO_2e Emissions in the Baseline Situation and the Scenarios

CED and CO_2e values, for both the baseline and the scenarios, are calculated for Austrian agriculture and displayed for each agricultural sector in Tables 3 and 4. In the scenarios, CED ranges from 30% to 38% lower than in the baseline situation, while CO_2e ranges from 30% to 37% lower. These headline statistics show the significant changes in energy demand and greenhouse gas emissions that would likely accompany a change to a healthier diet.

Table 3: CED in the baseline situation and the scenarios

	Baseline situation	Scenario self-sufficiency a	Scenario self-sufficiency b	Scenario self-sufficiency c	Scenario import/export a	Scenario import/export b	Scenario import/export c
	CED [MJ/capita]						
Crop cultivation	726	460	595	595	550	724	724
Grassland	460	143	152	220	152	169	198
Animal feed crop cultivation	363	319	325	320	290	290	290
Vegetable production	99	190	190	190	102	102	102
Fruit production	111	245	245	245	142	142	142
Animal husbandry	2,252	1,146	1,146	1,146	1,294	1,294	1,294
Sum	4,005	2,505	2,648	2,715	2,531	2,721	2,75
where the additional energy crop production is calculated as follows:							
Crop cultivation			135	0		174	0
Grassland			9	68		17	6
Sum without additional energy crop production			2,504	2,647		2,530	1,288

The data are based on the authors' own calculation

Fazeni and Steinmüller *Energy, Sustainability and Society* 2011 1:6, doi:10.1186/2192-0567-1-6

Table 4: CO_2e emissions in the baseline situation and the scenarios

	Baseline situation	Scenario self-sufficiency a	Scenario self-sufficiency b	Scenario self-sufficiency c	Scenario import/ export a	Scenario import/ export b	Scenario import/ export c
	CO2e [kg/per capita/annum]						
Crop cultivation	104	69	89	89	86	106	106
Grassland	118	39	42	61	42	47	53
Animal feed crop cultivation	62	56	56	56	50	50	50
Vegetable production	22	43	43	43	23	23	23
Fruit production	7	16	16	16	10	10	10
Animal husbandry	573	355	355	355	377	377	377
Sum	887	578	601	620	587	612	619

The data are based on the authors' own calculation.

Fazeni and Steinmüller *Energy, Sustainability and Society* 2011 1:6, doi: 10.1186/2192-0567-1-6

Although the CED of animal husbandry in the scenarios is nearly halved in comparison to the baseline situation, it remains the agricultural sector with the highest energy demand. Furthermore, these reductions are somewhat offset by a rise in energy demand from vegetable and fruit production, which would see an expansion in production area as a consequence of changed nutritional habits. Taken overall, the CED of Austrian agriculture shrinks in comparison to the baseline situation because less animal feed is needed. The CED of crop cultivation and grassland farming is lower in the scenario 'self-sufficiency a' than in the scenario 'import/export a' because of a difference in animal husbandry. In the scenario 'import/export a' there are more livestock to be fed due to the export of animal products. In sub-scenarios b and c, the CED of renewable energy feedstocks also needs to be included in the calculations, with sub-scenario c yielding a higher CED than b.

More specifically, the difference in CED between sub-scenarios a and b is due to the share of the CED derived from renewable energy feedstock production in sub-scenario b. In sub-scenario c, the use of grass from pasture as a renewable energy feedstock leads to a further increase in CED. An additional rise in crop cultivation in sub-scenario c is not possible because no more arable land is available.

The emission of CO_2e is closely connected with the CED of agriculture. Animal husbandry causes most of the CO_2e emissions of Austrian agriculture. Under the dietary change scenarios, CO_2e emissions fall reflecting an increased vegetable and fruit production and a decreased grassland farming and animal feed crop cultivation.

Renewable energy feedstock production leads to an additional CO_2e emission from agriculture in the sub-scenarios b and c. This additional CO_2e emission is the difference between the emissions in sub-scenarios a and b compared with b and c. Although renewable energy feedstocks are also produced on arable land in sub-scenario c, there is no increase in CO_2e emissions compared to scenario b because no further expansion of crop cultivation is possible.

Current research shows that Austrian agriculture would emit about 578 kg CO_2e per capita and year provided that nutrition is adapted to DGE recommendations. This discrepancy occurs because of taking the processing of foodstuffs into account [13]. It is difficult to compare the results from this research with other results due to differences in spatial and temporal system boundaries.

Production of Renewable Energy Based on Agricultural Raw Materials

In sub-scenario 'self-sufficiency c', the modeling projects 443,100 ha of renewable energy feedstock production, made up of 86,641 ha of arable land and 356,452 ha of grassland. The area of land used for renewable energy feedstock production in sub-scenario 'import/export c' is less than half of that used in sub-scenario self-sufficiency c, i.e., 208,800 ha, made up of 21,464 ha arable land and 187,360 ha of grassland. Looking at the outputs of the modeling, it is apparent that in practice, it would be all but impossible for Austrian agriculture to be self-sufficient in energy through the production of renewable energy feedstocks. However, a partial covering of CED is possible (Table 5).

Table 5: Contribution of renewable energy feedstock production to the CO_2e emissions of agriculture

	Scenario self-sufficiency b	Scenario self-sufficiency c	Scenario import/ export b	Scenario import/ export c
	CO2e [kg/per capita]			
Crop cultivation	20	0	21	0
Grassland	3	19	5	6

The data are based on the authors' own calculation.

Fazeni and Steinmüller *Energy, Sustainability and Society* 2011 1:6, doi: 10.1186/2192-0567-1-6

Table 6 illustrates that agriculture is able to make good a part of its CED by producing renewable feedstocks for energy production. In the best case (sub-scenario self-sufficiency c), enough energy is produced from renewable feedstocks to make good more than half of the entire agricultural CED. Determining factors in the level of CED replacement in agriculture are biofuel and biogas production. With diminished biodiesel production in the sub-scenario 'import/export b,' 21% of the CED can be made good by renewable energy feedstock production. In the sub-scenario import/export c, 37% of the entire CED can be made good. By contrast, in the sub-scenario self-sufficiency c, 68% of the CED is made good by renewable energy feedstock production. As much less bioethanol is produced in the scenarios 'import/export b/c,' total energy feedstock production in these scenarios, and therefore the extent to which CED is made good, is lower than in the case of the self-sufficiency scenarios. It should be pointed out that the data in Table 6 do not take into account the energy consumed in producing renewable energy feedstocks. Consequently, the values given for a share of CED made good are likely to overestimate the actual net level of replacement. Despite this, it is obvious that significant partial agricultural self-sufficiency in energy from renewable feedstocks is possible under the given conditions.

Table 6: Comparison of CED and energy production (self-sufficiency scenario and import/export scenario)

	Scenario self-sufficiency b	Scenario self-sufficiency c	Scenario import/export b	Scenario import/export c
Biodiesel [TJ]	2	2	5	5
Bioethanol [TJ]	4,862	4,862	1,366	1,366
Biogas [TJ]	3,415	10,035	3,203	6,773
Sum [TJ]	8,279	14,897	4,577	8,141
CED [TJ]	21,530	22,091	22,115	22,359
Proportion of CED made good [%]	38%	68%	21%	37%

The data are based on the authors' own calculation. TJ, terajoule.

Fazeni and Steinmüller *Energy, Sustainability and Society* 2011 1:6, doi: 10.1186/2192-0567-1-6

In the sub-scenario 'self-sufficiency b,' about 521,916 ha are used for food production, with a much larger area (1,520,710 ha) used for animal feed production. About 8% of the whole cultivated agricultural area is used for renewable energy feedstock production. The picture is similar in the sub-scenario import/export b, where 461,416 ha of land are used for food production and 1,949,839 ha are used for animal feed production. Only about 10% of the entire agricultural land employed in this sub-scenario is applied for renewable energy feedstock production.

The Direct Energy Demand of Agriculture

In self-sufficiency scenario, Austrian agriculture requires about 713 GWh of fuel, 815 GWh of thermal energy, and about 134 GWh of electricity per year. These results were derived by taking the direct energy requirements (per unit of the different crop and animal enterprises), multiplying these by the observed crop production areas and livestock numbers and aggregating to the national level [50]. For the sub-scenarios self-sufficiency b and import/export b, the target is that agriculture produces enough renewable energy feedstocks on free agricultural land to make it as close to self-sufficiency as possible in biodiesel as well as heat and electricity from biogas technology. In addition, enough feedstocks (wheat and maize) have to be cultivated by agriculture annually in order to utilize the capacity of Austria's agriculture and only bioethanol plants to the fullest.

The direct energy demand of agriculture in the import/export scenario is slightly lower than in the self-sufficiency scenario. This is because of the higher proportion of imported goods. So, in import/export scenario, agriculture needs about 755 GWh of fuel, 802 GWh of thermal energy, and about 130 GWh of electricity in total per year [50]. Various factors influence the amount of direct energy needed. The ratio of imported to domestically produced agricultural products has a significant impact on direct energy consumption. A larger share of imported vegetables implies a decrease in the thermal energy needed for cultivation under glass and a lower fuel demand for machinery. Additionally, higher exports of animal products cause an increase in fuel demand for crop cultivation because more animal feed has to be produced domestically.

Consequently, there is a supply gap of 105 GWh. As a result, agriculture cannot be self-sufficient in biodiesel in the sub-scenarios nor can the additive obligation of 5.75% to fossil fuels be fulfilled [51]. The situation is different in the import/export scenario: the ratio of imports to exports not only determines the direct energy consumption, but also influences land use and consequently crop rotation. As a result of decreased land use due to imports and changes in crop rotation, rape for biodiesel production is cultivated on 154,320 ha. The expansion of rape cultivation is attributable to the imports of oil seeds for human nutrition. Another important fact is the import of fish, which is an important supplier of omega-3 and omega-6 fatty acids. As a result, less oil seeds are needed to meet the fatty acid needs of the Austrian population [50]. This implies a biodiesel production of 1,512 GWh. Agriculture consumes only 755 GWh of biodiesel, and consequently, 757 GWh of biodiesel is available to fulfill the additive obligation or for other uses.

In the sub-scenario self-sufficiency b, 45,143 ha grassland and in sub-scenario import/export b, about 82,000 ha grassland are used for biogas production. In the sub-scenario self-sufficiency b, silage maize is used for biogas production in addition to grassland. In all, 10,393 ha for silage maize is available for biogas production. By contrast, no land is available for silage maize production in the import/export scenario; so, more grassland has to be assigned to the production of biogas. The difference in silage maize production between the two scenarios, self-sufficiency and import/export, again reveals the impact of importing and exporting agricultural goods in Austria. In the import/export scenario, the export of meat induces more animal husbandry so more land is needed for animal feed production, and given the crop rotation constraints, it is not possible to produce more silage maize in this scenario.

In the scenario self-sufficiency b, a total of 200,000 m^3 bioethanol is produced. In the import/export scenario, the production situation for bioethanol feedstocks differs; overall, maize is grown on 6,949 ha and wheat, on 14,515 ha for bioethanol production. In all, 64,522 m^3 are produced in the import/export scenario; so, the capacity of Austria's only bioethanol production plant is not used to the fullest. The increase in meat exports and in animal husbandry necessitates more animal feed production so less land is available for the production of wheat and maize as bioethanol feedstocks.

The only difference between the scenarios self-sufficiency b and self-sufficiency c and between the scenarios import/export b and import/export c is the full usage of grassland for biogas production. In the scenario self-sufficiency c, an additional of 356,452 ha of grassland is used for biogas production. A different situation is indicated in the scenario import/export c, in which the area of grassland for biogas production is lesser than in the scenario self-sufficiency c. In the scenario import/export c, a total of 192,444 ha grassland is available for biogas production. The area of grassland available for biogas production in the scenario import/export c is smaller because of the export of animal products and the simultaneous increase in animal husbandry so that more grass is needed for animal feed. The results of the various scenarios are shown in Figure 3 of this article.

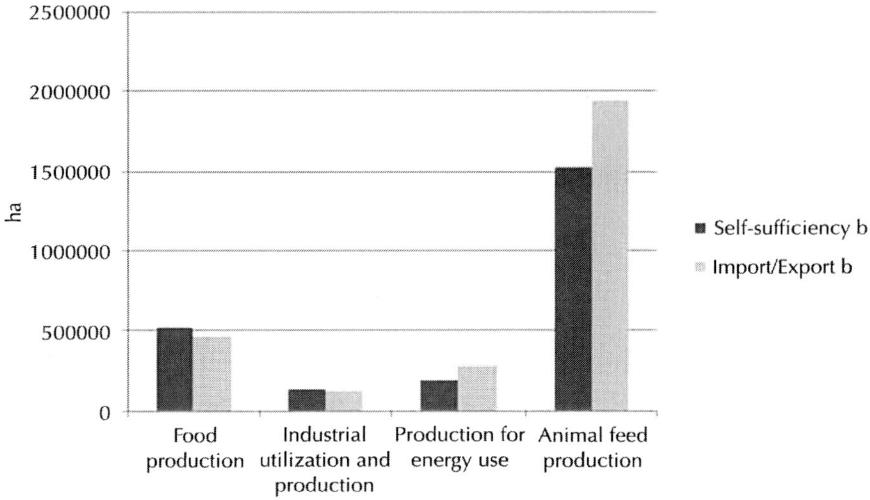

Figure 3: Agricultural land for different uses in the sub-scenarios self-sufficiency b and import/export b. The data are based on the authors' calculation.

DISCUSSION

This research has shown the extent to which the energy demand and greenhouse gas emissions of agriculture can be influenced by changes in human nutritional habits. A strong correlation between nutritional habits, resource demand, and the environmental burden of agriculture can be inferred. Although the study has Austrian agriculture as its particular focus, this correlation has already been shown in other studies with a different territorial focus [14, 15, and 52]. The results of the present study show that a decrease in meat consumption, arising from a change in diet, causes a release of arable land. This would be a significant outcome for Austrian agriculture with its current dominance by livestock production, driven by high rates of meat consumption both in Austria and its trading partners. These results confirm the findings of other research carried out internationally [14-20, 52].

It is important to examine the correlation of nutritional habits with agricultural energy demand and greenhouse gas emissions at a regional level because specific production methods and circumstances can then be taken into account. The main aim of this study was to examine how a change in diet (and concomitant release of land for renewable energy feedstock production) influences the CED and CO_2e emissions of Austrian agriculture. To do this, Austrian agricultural production was modeled as a single average farm, where all agricultural goods in demand are produced. Applying this method involves some uncertainties because some parameters cannot be determined in detail. As a result, no statements about soil quality and soil management methods are made. As soil management influences emissions from agriculture and the demand for energy, a detailed scenario calculation for each Austrian production area would lead to results different from the 'averages' presented in this study. In some cases, the values for emissions and energy demand would be higher, for example, in intensive production areas; in other cases, they would be lower, for example, in extensive production areas. Another limitation of this study is that, for energy crop production on redundant land, no precise statements can be made about where this production is located and whether this land is in fact suitable for energy production, or even whether it would be economic to convert surplus land to these uses. Sustainable economic activity by farmers may not lead to the release of land where there are no profitable alternative uses; under these circumstances, land is likely to remain in livestock production, albeit under more extensive conditions. It is therefore a simplifying assumption of the modeling that land that is surplus to food and feed production must be diverted to renewable energy crop production and to only these uses.

Other limiting factors can be identified in the CED calculation. Agriculture receives no energy or emission credits in the sense of the LCA methodology according to ISO 14040 for producing energy crops. As a result, the emissions and energy demand of agriculture are slightly overestimated because emission and energy credits would lower the values of these parameters [52]. Regarding the energy consumption of agriculture, the aim was to examine the demand side; so, the energy input to and output from agriculture are not compared. A further change in energy demand and emissions can be induced if agricultural emissions and energy consumption abroad are calculated. System boundaries have to be set so as to reduce the amount of data that has to be analyzed to manageable proportions. This should not be taken to mean that energy demand and emission output from Austrian agriculture can be brought to zero by simply importing all goods.

BRAINBOWS estimates for Austria that in the year 2020 about 455,000 ha agricultural land could be used for renewable energy crop production [53]. This is slightly higher than the estimate in the sub-scenario self-sufficiency c, where 443,000 ha are projected to be available for renewable energy crop production. It is questionable whether the potential estimated in the study by BRAINBOWS [53] is realistic because this estimate is based on the assumption that set-aside land is used, that surplus goods which are exported at the moment are used domestically, and that demand for animal feed goes down because of the use of co-products from food and particularly biofuel processing. In addition, the development of higher yielding crops and the use of catch crops should guarantee that this potential is realized by 2020. Even if, under the assumptions made in the study by BRAINBOWS [53], a similar amount of agricultural land can be used for energy crops, a change in diet generates further potential. Another advantage of land released because of a change in diet is that this land does not compete with food production.

CONCLUSIONS

The present work has shown that change in nutritional habits can have a great influence on agricultural energy consumption and greenhouse gas emissions. Above all, eating less meat would lead to a decrease in negative agricultural environmental impacts. This research involves some uncertainties caused by the simplifications necessarily involved with treating Austrian agriculture as a single 'average' farm. As a result, it was not possible to consider the different conditions of production specific to various farming regions. Despite these uncertainties, the positive effects of reducing meat consumption and basing nutrition on plants to a greater extent on the agricultural energy and emission balance are obvious from the modeling and well attested in the literature. Furthermore, changed nutritional habits can contribute to the achievement of policy targets defined for renewable energy use through the release of redundant land, where a large part of which can be used for renewable energy crops. So, the solution to the problem of increased competition for land for bioenergy production might well be not to increase the area under cultivation in sensitive regions, not to plow up grassland for crop cultivation, nor to increase the agricultural output by applying more pesticides and fertilizers. Even under consistent agricultural production methods in Austria, changed nutritional habits make more arable land available for renewable energy crops. As a consequence, changing nutritional habits would be desirable not only because of the potential benefits that might be obtained in terms of human health, but also because of these secondary emissions and renewable energy benefits.

The novelty of this work is that the impacts of dietary choices on the availability of land for renewable energy production and the positive CED and emissions benefits are examined simultaneously. Existing studies on this topic often focus on the impact of dietary choices either on energy and emissions or on the availability of land, e.g., in the studies conducted by Carlsson-Kanyama [4], Eshel and Martin [9], Risku-Norja et al. [11], Gerbens-Leenes and Nonhebel [54], Elferink and Nonhebel [23], and Dale et al. [52]. This study merges these approaches. The results of this analysis suggest that new options for mitigating greenhouse gas emissions and reducing the use of fossil energy are feasible. A change in diet would be the first step to a more sustainable agriculture and more sustainable production of renewable energy crops. Thus, this work also demonstrates the importance of an integrated policy design, encapsulating nutrition, agriculture and renewable energy.

The assumption on arable land and grassland available for renewable energy feedstock production in the examined scenario involves an expansion compared to the baseline situation. On the other hand the scenario estimates are lower than the estimates presented in the Austrian Biomass Action Plan. In particular, the import/export scenario shows more modest results than the Biomass Action Plan. By contrast, the self-sufficiency scenario shows results quite similar to the potential estimated in a biomass resource potential study for Austria [52]. For purposes of comparison with existing studies of biomass potential in Austria, the scenario results for maximum renewable energy feedstock production were chosen. In future the role of energy production from agricultural residues will be strengthened and therefore the renewable energy production potential will increase further [55].

However, with a maximum of about 8% of agricultural land used for renewable energy crops under any scenario, the results of the study also show that most of the greater part of agricultural land will always be needed for food and feed production, even if we assume the most positive outcomes in terms of changed nutritional habits.

AUTHORS' CONTRIBUTIONS

KF carried out the CED and CO_2e calculations as well as the calculation of the energy produced on the redundant land, wrote the manuscript, and was responsible for textual design of the paper. HS contributed to the underlying assumptions and the scenario definition. Furthermore, HS proofread the manuscript and gave some important evidences concerning the structure and content of the paper. All authors read and approved the final manuscript.

ACKNOWLEDGEMENTS

This publication has evolved from a project within the proVISION program, funded by the Austrian Federal Ministry of Science and Research. ProVISION is aimed at implementing Austria's FORNE strategy (research for sustainable development) together with complementary research programs, creating the scientific basis for the country's sustainability strategy.

REFERENCES

1. Steinfeld H, Gerber P, Wassenaar T, Castel V, Rosales M, de Haan C (2006) Livestock's long shadow: environmental issues and options. FAO, Rome.
2. Mummey D, Smith J, Bluhm G (1998) Assessment of alternative soil management practices on N_2O emissions from US agriculture. Agr Ecosyst Environ 70:79-87
3. Smith P, Martino D, Cai Z, Gwary D, Janzen H, Kumar P, McCarl B, Ogle S, O'Mara F, Rice C, Scholes B, Sirotenko O, Howden M, MacAllister T, Pan G, Romanenkov V, Schneider U, Towprayoon

S (2007) Policy and technological constraints to implementation of greenhouse gas mitigation options in agriculture. Agr Ecosyst Environ 118:6-28

4. Carlsson-Kanyama A (1998) Climate change and dietary choices-how can emissions of greenhouse gases from food consumption be reduced? Food Pol 23:277-293
5. Kramer KJ, Moll HC, Nonhebel S, Wilting HC (1999) Greenhouse gas emissions related to Dutch food consumption. Energy Pol 27:203-216
6. Pimentel D, Pimentel M (2003) Sustainability of meat-based and plant-based diets and the environment. Am J Clin Nutr 78:
7. Reijinders L, Soret S (2003) Quantification of the environmental impact of different dietary protein choices. Am J ClinNutr 78:
8. Wallen A, Brandt N, Wennersten R (2004) Does Swedish consumer's choice of food influence greenhouse gas emissions? Environ Sci Pol 7:525-535
9. Eshel G, Martin P (2006) Diet, energy and global warming. Earth Interact 10:1-16
10. Weber Ch, Matthews HS (2008) Food-miles and the relative climate impacts of food choices in the United States. Envrion Sci Technol 42:3508-3513
11. Risku-Norja H, Kurppa S, Helenius J (2009) Impact of consumers' diet choices on greenhouse gas emissions. In: Koskela M, Vinnari M (eds) Future of the consumer society, Writers & Finland Futures Research Center, Tampere. pp 159-17
12. Stehfest E, Bouwman L, van Vuuren D, den Elzen M, Eickhout B, Kabat P (2009) Climate benefits of changing diet. Climatic Change 95:83-102
13. Freyer B, Weik S (2008) Impact of different agricultural systems and patterns of consumption on greenhouse-gas emissions in Austria. 16th IFOAM Organic World Congress, Modena. 16-20 June 2008
14. Garnett T (2009) Livestock-related greenhouse gas emissions: impacts and options for policy makers. Environ Sci Technol 12:491-503
15. Popp A, Lotze-Gampen H, Bodirsky B (2010) Food consumption, diet shifts and associated non-CO_2 greenhouse gases from agricultural production. Global Environ Change 20:451-462

16. McMichael AJ, Powles JW, Butler CD, Uauy R (2007) Food, livestock production, energy, climate change, and health. Lancet 370:1253-1263
17. Keyzer MA, Merbis MD, Pawel IFPW, van Wesenbeeck CFA (2005) Diet shifts towards meat and the effects on cereal use: can we feed the animals in 2030? Ecol Econ 55:187-202
18. Rathmann R, Szklo A, Schaeffer R (2010) Land use competition for production of food and liquid biofuels: an analysis of the arguments in the current debate. Renew Energy 35:14-22
19. Karp A, Richter GM (2011) Meeting the challenge of food and energy security. J Exp Bot. 1-9
20. Babcock BA (2008) Breaking the link between food and biofuels. In: Briefing Paper 08-BP 53. Center for Agricultural and Rural Development, Iowa State University.
21. Rosegrant MW (2008) Biofuels and grain prices. International Food and Policy Institute.
22. Young AL (2009) Finding the balance between food and biofuels. Environ Sci Pol Res 16:117-119
23. Elferink EV, Nonhebel S (2007) Variations in requirements for meat production. J Clean Prod 15:1778-1786
24. Wallén A, Brandt N, Wennersten R (2004) Does Swedish consumers' choice of food influence greenhouse gas emissions? Environ Sci Pol 7:525-535
25. Nonhebel S (2007) Energy from agricultural residues and consequences for land requirements for food production. Agr Syst 94:586-592
26. Owens JW (1996) Life-cycle assessment in relation to risk assessment: an evolving perspective. Risk Anal 17:359-365
27. Jones MR (1989) Analysis of the use of energy in agriculture-approaches and problems. Agr Syst 29:339-355
28. Hutter C, Koehler D (1999) Ökobilanzierung mit Hilfe der KEA-Datenbank. Forschungsstelle für Energiewirtschaft, München.
29. Kloepffer W (1997) In Defense of the cumulative energy demand. Int J LCA 2:61
30. Seebacher U, Oehme I, Suscheck- Berger J, Winds perger A, Steinlechner S (2003) PUIS- Produktbezogene

Umweltinformationssysteme in österreichischen Unternehmen. BMVIT, Wien.

31. Fischer J (1999) Energy inputs in Swiss agriculture. FAT. Working Paper. 99-01
32. Biedermann G (2009) Kumulierter Energieaufwand (KEA) der Weizenproduktion bei verschiedenen Produktionssystemen (konventionell und ökologisch) und verschiedenen Bodenbearbeitungssystemen (Pflug, Mulchsaat, Direktsaat). In: Master›s Thesis. University of Natural Resources and Life Science Vienna.
33. Payraudeau S, van der Werf HMG (2005) Environmental impact assessment for a farming region: a review of methods. Agr Ecosyst Environ 107:1-19
34. Olesen JE, Schelde K, Weiske A, Weisbjerg MR, Asman WAH, Djurhuus (2006) Modelling greenhouse gas emissions from European conventional and organic dairy farms. Agr Ecosyst Environ 112:207-220
35. Bentrup F, Küsters J, Kuhlmann H, Lammel J (2004) Environmental impact assessment of agricultural production systems using the life cycle assessment methodology. I. Theoretical concept of a LCA method tailored to crop production. Eur J Agron 20:247-264
36. Haas G, Wetterich F, Geier U (2000) Life cycle assessment framework in agriculture on the farm level. Int J LCA 5:345-348
37. Roy P, Nei D, Orikasa T, Xu Q, Okadome H, Nakamura N, Shiina T (2009) A review of life cycle assessment (LCA) on some food products. J Food Eng 90:1-10
38. Hülsbergen KJ, Feil B, Biermann S, Rathke GW, Kalk WD, Diepenbrock W (2001) A method of energy balancing in crop production and its application in a long-term fertilizer trial. Agr Ecosyst Environ 86:303-321
39. Jones MR (1989) Analysis of the use of energy in agriculture-approaches and problems. Agr Syst 29:339-355
40. Institut für Angewandte Ökologie e.V (2008) Globales Emission Modell Integrierter Systeme (GEMIS).
41. KTBL (2008) KTBL-Datensammlung Betriebsplanung 2008/09.

42. OEKL (2009) Treibstoffverbrauch in der Land- und Forstwirtschaft 2009.
43. Demerci M (2001) Ermittlung der Deckungsbeiträge der wichtigsten Gemüsekulturen im Gewächshaus in Österreich. In: PhD Thesis. University of Natural Resources and Life Science Vienna.
44. Statistic Austria (2005) Garten-, Feldgemüsebau. [http:/ / www.statistik.at/ web_de/ statistiken/ land_und_forstwirtschaft/ agrarstruktur_flaechen_ertraege/ gartenbau_feldgemueseanbau/ index.html] Accessed 10 Sept 2010
45. Statistik Austria (2010) Energiegesamtrechnung. [http:/ / www.statistik.at/ web_de/ statistiken/ energie_und_umwelt/ energie/ energiegesamtrechnung/ index.html] Accessed 10 Sept 2010
46. Statistik Austria (2010) Energieeinsatz der Haushalte. [http:/ / www.statistik.at/ web_de/ statistiken/ energie_und_umwelt/ energie/ energieeinsatz_der_haushalte/ index.html] Accessed 10 Sept 2010
47. Zessner M, Steimueller H, Wagner KH, Krachler MM, Thaler S, Fazeni K, Helmich K, Weigl M, Ruzicka K, Heigl M, Kroiss H (2011) Gesunde Ernährung und Nachhaltigkeits-Grundlagen, Methodik und Erkenntnisse eines Forschungsprojektes im Rahmen des proVision Programmes des BMWF. ÖWAW. 5-6
48. Zessner M, Helmich K, Thaler S, Weigl M, Wagner KH, Haider T, Mayer MM, Heigl S (2011) Ernährung und Flächennutzung in Österreich. ÖWAW. 5-6 forthcoming
49. European Parliament, European Council (2009) Directive 2009/28/EC of the Parliament and of the Council of 23 April 2009 on the promotion of the use of energy from renewable sources and amending and subsequently repealing Directives 2001/77/ EC and 2003/30/EC. Brussels.
50. European Parliament, European Council (2003) Directive 2003/30/EC of the Parliament and of the Council of 8 May 2003 on the promotion of the use of biofuels or other renewable fuels for transport, Brussels.
51. Steinmueller H, Fazeni K (2011) Energiebilanzen der österreichischen Landwirtschaft unter Berücksichtigung von Ernährungsgewohnheiten. ÖWAW. 5-6

52. Dale BE, Bals BD, Kim S, Eranki P (2010) Biofuels done right: land efficient animal feeds enable large environmental and energy benefits. Environ Sci Technol 44:8385-8389
53. BRAINBOWS (2007) Biomasse-Ressourcenpotential in Österreich. Studie im Auftrag der Renergie Raffeisen Managementgesellschaft für erneuerbare Energie GmbH.
54. Gerbens-Leenes PW, Nonhebel S (2002) Consumption patterns and their effects on land required for food. Ecol Econ 42:185-199
55. Nonhebel S (2007) Energy from agricultural residues and consequences for land requirements for food production. Agr Syst 94:586-592

Chapter 3

Legal Protection Assessment of Different Inland Wetlands in Chile

Patricia Möller[1,2] and Andrés Muñoz-Pedreros[3]

[1]Facultad de Ciencias, Universidad Austral de Chile, Casilla 567, Valdivia, Chile

[2]Programa de Humedales, Centro de Estudios Agrarios y Ambientales, Casilla 164, Valdivia, Chile

[3]Núcleo de Estudios Ambientales NEA, Escuela de Ciencias Ambientales, Facultad de Recursos Naturales, Universidad Católica de Temuco, Casilla 15-D, Temuco, Chile

ABSTRACT

Background

Inland wetlands are well represented ecosystems in Chile that are subjected to various pressures affecting conservation. Protection means legal and administrative initiatives which promote the protection and/

or preservation of a wetland, either in its entirety, considering their areas of influence or its components.

Results

The aim of this work is to develop a methodology for estimating the value of protection of different types of inland wetlands in Chile. For this purpose: a) the Chilean regulations in relation to the issues of biodiversity, wetlands and water resources were compiled; (b) such legislation and its application were analyzed, (c) protection and restriction values of each legal standard was estimated, and then total protection value of standards applied to six types of wetlands in study. 47 legal rules related to protection of inland wetlands and eight directly or indirectly affecting conservation and wise use of wetlands were identified.

Conclusions

In Chile there is no specific statutory rules or regulations on wetlands. Current legal standards do not protect equally the different types of inland wetlands, being swamp forests, peatlands and brackish Andean lakes less protected. To improve wetlands conservation, incentives promoting wetlands destruction must be eliminated and promulgate specific regulations for proper management and conservation.

BACKGROUND

Chile is highly diverse in limnic systems mostly recognized as wetlands and defined as 'areas of marsh, fen, peatland or water, whether natural or artificial, permanent or temporary, static or flowing, fresh, brackish or salt waters, including areas of marine water where tide depth does not exceed the six meters' (Ramsar [2000]). This long definition includes very heterogeneous areas, such as swamps, rivers, lakes, seashores, and others (Mitsch and Gosselink [2000]). Wetlands are highly productive ecosystems (Novitzki et al. [1996]) characterized by multiple roles, e.g. hydrological, biogeochemical, habitat conservation, and food webs (Woodward and Wui [2001]), as well as providing goods and services relevant to human society (Barbier et al. [1997]). In Chile, wetlands are estimated to cover 4,498,060, 7 ha equivalent to 5.9% of national territory (CONAF/CONAMA [1997]).

Wetlands support high biodiversity and are presently recognized as the most threatened systems by human activities (Marín et al. [2006]). It affects various types of wetlands in Chile (Muñoz-Pedreros [2004], Peña-Cortés et al. [2006], Zegers et al. [2006], Figueroa et al. [2007]). Particularly, freshwater aquatic fauna conservation is mostly affected, such as fishes (Vila et al.[2006]), amphibians (Díaz-Paéz and Ortiz [2003], Veloso [2006]), molluscs (Valdovinos et al.[2005]), and decapod crustaceans (Bahamonde et al. [1998], Pérez-Losada et al. [2002]). Inland wetland conservation is a global priority (Abell [2002], Dudley [2008]). However, their particular territorial location and the difficulty to apply current protection categories make a difficult management as protected areas. As traditional ecosystem conservation methods are not well implemented in inland aquatic environments, different conceptual approaches are suggested (e.g. freshwater focal area, critical management zone, and catchment management zones (Abell et al.[2007])).

Governance concept applied to natural resources, mainly in waters, has become important (Iza and Rovere [2006]). Governance is understood as the economic, political, and administrative practice to manage every country affair. This includes mechanisms, processes, and institutions by which citizens express their interests, exercise their rights, meet their requirements, and mediate their differences. In this context, conservation of wetlands should be addressed under various environmental management tools as a permanent process, where various stakeholders and public, private, and civil society develop specific efforts to preserve, maintain, restore, and make a sustainable use of environment. Environmental management uses different and diverse origin instruments, which can be classified in four main categories (sensu Rodríguez-Becerra and Espinoza [2002]): direct regulation, administration, economics, and education (including research, technical assistance, and environmental information). Direct regulatory tools which are also known as command and control regulations prevail in the environmental management and consist of mandatory regulations and standards, which establish environmental quality targets as well as management and preservation of renewable natural resources of environment. Legislation creates legal tools and standards to comply with principles and reach the aims (Asenjo [2006]).

Brañes ([2000]) defines the environmental regulation as a set of standards dealing with legal protection of those conditions making every life to be possible, considering for this purpose relationships among many biotic and abiotic elements in the environment, such as system or ecosystem. To Fernández ([2004]), the environmental law gathers standards, regulations, and principles recognized as legally protected, safeguarding environmental systems, in a global and inclusive perspective that differ from merely legislation of environmental incidence.

It may be understood here that legal protection of wetlands is provided by all legal and administrative initiatives aimed to protect and/or preserve them. There are many legal standards that apply to inland wetlands in Chile, many of them are sectoral standards related to their components, ecological functions, and biodiversity, which would favor or regulate their status and permanence. Selection and analysis of these regulations will allow to provide foundations, which make possible assessment of legal degree protection that they provide to these systems and to integrity as ecological systems. In Chile, regulations addressed to the conservation of water resources as a whole have been mainly aimed to develop economic activities related to the exploitation of a natural resource. Until 1994, prior to the Environmental Basis Act (Act No. 19,300) enactment, there were only sector legal standards, without global environmental protection objective, where legal rules were mainly aimed to protect health of human life and only incidental protection of nature (Olivares [2010]). Act No. 19,300 started regulation process of environmental standards, as well as the creation of institutions providing the state of management tools in this area (Rojas [2011]).

However, Hermosilla ([2004]) estimates that while keeping current protection system and guarantees of property rights in Chile over the common well, effective protection to natural systems including wetland ecosystems will not be possible.

Prior this research, there was no compilation and analysis of rules applied to wetland systems which allows to determine real legal protection in taking care of conservation, but there are legal standards which should be assessed under this context, like those regulations enacted which damage preservation and rational use of inland wetlands. The rational use concept is defined by the Ramsar

Convention as the maintenance of ecological features, achieved through the implementation of ecosystem-based approach, within sustainable development context.

Under the Chilean legal system, environmental heritage conservation is the rational use and exploitation or repair if any, of environment components, to assure sustainability and regeneration capacity (Act No. 19,300; Article 2b); however, it does not point out which are those components.

Some environmental components, those considered to be relevant issues for this study among them, are noted by Fernández ([2004]): a) land or sea waters, surface water or groundwater, streams or standing waters; b) land, soil and subsoil, including beds, bottom and subsoil of ground waters; c) flora and wildlife, land or water; d) microflora and fauna of land, soil and subsoil, streams or water bodies and beds, bottoms and subsoil of these streams or water bodies; e) genetic diversity and patterns and factors regulating flow; f) natural scenic beauty and rural or urban landscape; g) essential ecological processes. These basic components of the environment can be damaged when misused which results in extinction or serious damage, prevents regeneration, and causes environmental damage determined in law as 'any loss, decrease, impairment, or significant impairment associated to environment or to one or more of its components' (Act No. 19,300: Art. 2e).

Some of the elements or factors which could damage or degrade the environment are (to Fernández [2004]): a) any kind of pollution; b) erosion, salinization, alkalinization, infestation, flooding, sedimentation, and desertification of soil and land; c) logging or unreasonable and uncontrolled destruction of trees and shrubs as well as extractive forest use and other vegetation destruction; d) monocultures, overgrazing, and, as a whole, any cultural practice with harmful effects on the environment; e) sedimentation of water streams and lakes; f) harmful alterations of natural water flows; g) adverse changes and misuse of water beds; h) wild flora and fauna over-exploitation; i) elimination, destruction, or degradation of endangered species from flora and fauna habitat; j) wetland eutrophication origins; k) introduction or distribution of exotic plants or animals coming from a national biogeographical province different to natural; l) introduction or spread of animal diseases or plant pests; m) use of non-biodegradable products or substances; n) the accumulation of waste or inappropriate waste disposal; o) visual landscape destruction or alteration, and, as a whole, any act or omission affecting negatively the basic composition, behavior, and natural potential of environment; threatening land genetic viability or affecting life, health, integrity, or development of man, plants, or animals. These environmental components and environmental degradation factors were considered in the analysis of rules to determine protection concept granted by the Chilean law to the rational use of wetlands. Thus, the aim of this work is to develop a multicriteria methodology to quantify the level of legal protection that different types of inland wetlands have in Chile.

METHODS

Compilation and Analysis of Current Standards

Compilation and analysis of current legislation related to wetlands and water resources, including legal standards with description and characterization, was carried out, consulting different sources of reference (e.g., CONAMA [1996a], [1996b], [1997a], [1997b], [2009], Castillo [1994], Gallardo[1985] Hermosilla [2004], Ortiz [1986], [1990], Valenzuela [1994], Fernández [2004], Bravo[2010]). Also, the website of the Library of Congress of Chile was reviewed (www. leychile.cl*webcite*), which has a compendium of environmental legal standards and acts classified by topics. The review covered Chilean regulations until April 2012.

The identification of the corresponding legal standards was based on the interest of this study (e.g., fresh water fauna, water resources, wetlands, watersheds), selecting standards aimed at protecting and preserving aspects and components involved in direct or indirect conservation of inland wetlands. Legal standards in force which restrict conservation and protection of inland wetlands, as well as issues related to functionality and biodiversity, were also identified. Selected rules were hierarchically ordered according to Gallardo ([1985]).

Multicriteria Method for Legal Protection Assessment

Wetland legal protection will tend to protect and/or conserve, either entirely, in components or considering the influence areas. This protection was estimated with a multicriteria methodology developed according to the following procedure (Figure 1): (a) compilation and analysis of existing legislation related to inland wetland issues (aspects related to biodiversity conservation and rational use included); also, current regulations which restrict conservation and rational use of inland wetlands were identified; (b) estimation of protection value (PV) provided by each legal regulation that considers applicability and standard hierarchy; (c) estimation of restriction value (RV) provided by each legal regulation instrument that also considers applicability and legal standard hierarchy; and (d) finally, estimation of real protection value (RPV) by summing protection values (PVs) of each legal standard applied to wetland of interest and restriction values (RVs) applied to the same type of wetland.

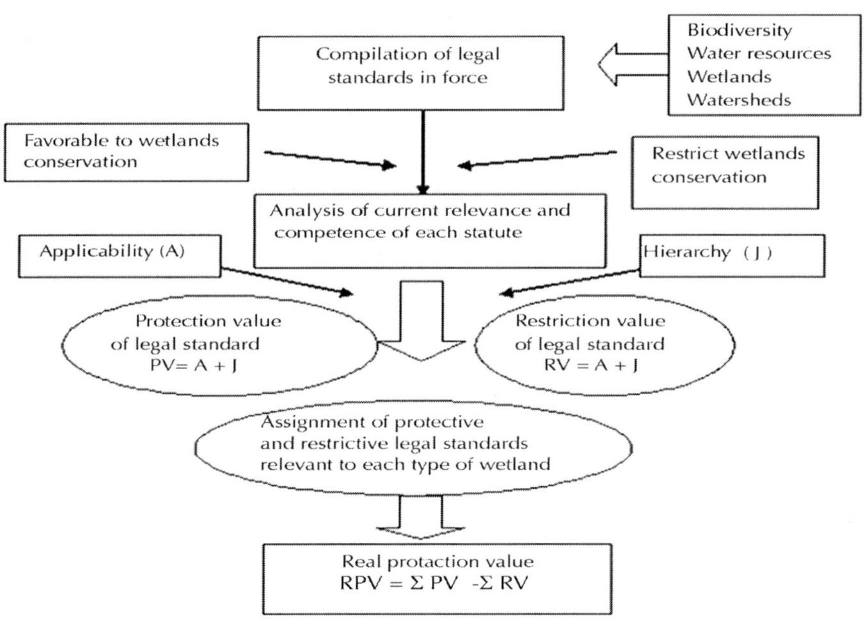

Figure 1: Methodological flow diagram to estimate value of legal protection for wetlands in Chile.

Protection Value (PV) Calculation

Protection value considers applicability and legal standard hierarchy (Figure 1). Applicability (A) is given as expressly enacted for wetland conservation purposes (value 3); or if secondarily, it could be considered as it refers to conservation of some environmental components emphasizing the aquatic system as a natural system (value 2), or if it is only tangential or generically involved with conservation of any component without reference to the natural system (value 1). Legal standards applicability was assessed, considering the reference which protects or preserves. It was classified in three levels: a) wetland system, including reference to 'wetlands' 'aquatic environment', 'aquatic ecosystems', or 'biodiversity' concepts (in ecosystem level); b) components, e.g., water, flora, and plants, with reference to aquatic system (e.g., river or lake); and c) component, e.g., water, wildlife, flora, or vegetation, without reference to aquatic system. These three levels were assigned with values 3, 2, and 1 respectively, getting applicability of each of the legal regulations. Hierarchy (H) was assessed by assigning the maximum value (3) to legal regulations, an average hierarchy (value 2) to decrees with force of law and law decrees, and lower hierarchy (value 1) to standard rules (e.g., regulations, simple decrees, instructions, supreme decrees, resolutions). For details of this hierarchy, see.

Legal protection estimated value (PV) was calculated using the following formula: $PV = A + J$, where $PV =$ legal protection value of each legal rule, $A =$ applicability of standard to each wetland type, and $J =$ hierarchy that the standard has in the law of Chilean legal system. Thus, the protection value (PV) of each legal instrument is deployed between 2 ($PV = 1 + 1$) (minimum protection) and 6 ($PV = 3 + 3$) (maximum protection). The sum of the variables considers that the weight of each is the same.

Restriction Value (RV) Calculation

It is estimated by analyzing protection restrictions of legal standards and the rational use of diverse wetlands. Applicability and legal hierarchy standards (Figure 1) were also considered. Applicability (A) was determined, considering what and how does it affect wetlands. The implementation effects are the complete wetland system destruction or any structural component (value 3); collateral effect (value 2) and whether it may only eventually affect (value 1). Hierarchy (H) was valued in same way as the protection value. Restriction value (RV) was calculated as: $RV = A + J$, where RV = restriction value of each legal rule, A = statute applicability, and J = standard hierarchy. Thus, restriction value (RV) to each legal regulation is deployed between 2 ($VR = 1 + 1$) (minimum constraint) and 6 ($RV = 3 + 3$) (maximum constraint). The sum of variables considers the weight of each one to be the same. Integration of parameters included and their subsequent validation was done by a team of 11 experts using the method of Delphi (Linston and Turoff[1975]). The group of experts included professionals from different disciplines, with over five years of professional experience, who in spite of not having legal formation, they were linked to these subjects. Criteria considered specialists in aquatic resources (4), relation with public institutions (3), and environmental management experience (5) Work methodologies were workshops, documents, and questionnaire previously associated, which were duly submitted to experts to review and compile observations by using templates. Observations from experts were systematized and incorporated to a document which was forwarded for assent repeating procedure described to obtain consensus. The algorithm was applied to different types of inland wetlands of Chile.

Wetlands Studied

Six types of wetlands were selected based on representativeness and wetland they represented. Representation considered frequency, defined as abundance, in a survey of 1,215 inland wetlands (CEA/FIP [2010]), with Scott and Carbonell records ([1986]); Schlatter et al. ([2001]); and Lopez-Lanus and White ([2005]), which was refined and grouped according to the type by Dugan ([1992]) and considerations by Ramsar ([2000]) and Ramirez et al. ([2002]). Based on the above, selected wetlands correspond to fluvial systems, lakes, and palustrine areas, which are: (a) rivers, (b) lagoons, (c) brackish Andean lakes, (d) high-altitude Andean peatlands called 'vegas' and 'bofedales', (e) freshwater swamp forests, and (f) permanent freshwater marshes and swamps also called locally as 'bañados' and 'ciénagas'.

RESULTS

Legal Standards for Wetland Protection

Standards were selected either since they address wetland as entirety natural system or they refer to some constituting component, such as water (in quality), aquatic life, wildlife, and others, as well as the environmental and aesthetic roles. Same way, those referred to any basin component with impact on aspects above mentioned (conservation of watershed vegetation, woodland springs, and riparian zones) were also included. Those standards which currently or potentially damage wetlands or any of their components were also included. Rules were considered from current or potential application point of view. Forty-seven legal rules directly or indirectly related to the protection of wetlands were identified.

Protection Value (PV) of Legal Standards

Each legal standard protection value is shown in Table 1, which presents 47 legal rules with each potential protection value, according to applicability and hierarchy. It can be observed that they mostly (30 standards) have protection values 2 or 3, it means low; and 11 standards have intermediate protection value (value 4) while only six standards have higher protection values (values 5 and 6). The total protection value given by adding all protection standards is 144.

Table 1: Applicability (A) of legal standards affecting Chile's inland wetlands depending on what it protects or preserves, hierarchy (J) and protection value (PV) of each legal rule, as well as it total value protection

Legal standard	What protects/ preserves?	A	J	PV
Act N 19,300	Environment	3	3	6
DS N° 531 Washington Convention	Flora and fauna in protected area systems	1	1	2

DS N 141 CITES	Components (flora and fauna)	1	1	2
DL N° 3,485 Ramsar Convention	Systems (wetland)	3	2	5
DS N° 868 Migration of species	Component (fauna)	1	1	2
Decree N° 1,963 Convention in biodiversity	Biological diversity	3	1	4
Act N° 11,402 Barrier works and standardization of bank and channel	Reforestation river basin (headwaters)	2	3	5
Act N° 17,288 National Monuments	Sites (Research interests)	1	3	4
Act N° 18,892 Fishery and Agricultural Standards	Components (fauna, such as hydro-biological resources (fisheries and aquaculture))	2	3	5
Act N° 18,902 Sanitary health superintendency creation; liquid industrial residues control	Component (water quality for human purposes)	1	3	4
Act N° 19,473 Hunting Law	Components (fauna)	1	3	4
DFL N° 208 Provisions for development of fishing activities	Components (water quality, fish, and seafood)	1	2	3
DFL N° 701 About forest development	Components (soil for quality purposes/water quantity for irrigation and drinking)	1	2	3

Act N° 20.283 Native Forest Recovery and Forestry Development rule	Components (native forest associated to springs, natural bodies, and water courses)	3	3	6
Act N° 20,256 Recreational fishing standards	Ecosystem	3	3	6
Act N° 20,411 Prevents constitution of water rights	Components (aquifer)	1	3	4
DFL N° 725 Sanitary Health Legal Standards Art. 73	Components (quality/water quantity and human health)	1	2	3
DFL N° 1,122 Water Legal Standards	Components (water supply quantity for human use)	2	2	4
DS N° 1 Regulations for water pollution control	Components (water supply quantity for human use)	1	1	2
DS N° 5 Hunting law approval	Components (fauna)	1	1	2
DS Nv° 29 Approval of classification standards according to wild species conservation conditions	Components (wild species)	1	1	2
DS N° 30 Law 19,300	Systems (wetland)	3	1	4
DS N° 46 Groundwater emission standards	Components (water supply quantity for human use)	1	1	2

DS N° 33 Classification process of species according to conservation condition	Components (wild species)	1	1	2
DS N° 41 Classification process of species according to conservation condition	Components (wild species)	1	1	2
DS N° 42 Classification process of species according to conservation condition	Components (wild species)	1	1	2
DS N° 51 Classification process of species according to conservation condition	Components (ichthyofauna species)	1	1	2
DS N° 82 Soil, water, and wetlands legal standards Act N 20,283	Systems (bodies and natural waterflows)	3	1	4
DS N° 90 Standards for liquid residues discharge into surface freshwater	Components (water quality in four types of use)	1	1	2

DS N° 95 Environmental Quality Standards Law	Components (water quality for human and environmental health)	2	1	3
DS N° 93 (2009) Native Forest Recovery Law	Components (native forest with soil, water, and wetland protection areas)	3	1	4
DS N° 193 Law Decree 701 General rules	Components (soil, courses and water bodies, flora and fauna)	1	1	2
DS N° 210 Reforestation and recreational fishing standards	Components (ichthyofauna)	1	1	2
DS N° 236 Standards for disposals not discharged to sewage network	Components (water quality for human consumption)	1	1	2
DS N° 238 Park and marine reserves standards (including inland waters)	Systems (wetland)	3	1	4
DS N° 320 Aquaculture environmental standards	Systems (river and lake)	2	1	3
DS N° 430 Reviewed document for fisheries and aquaculture Law	Systems (river and lake)	2	1	3
DS N° 594 Health and work place environmental condition standards	Components (groundwaters and surface waters)	1	1	2

DS N° 351 Industrial liquid residues discharge depuration and neutralization standards	Components (groundwaters and surface waters)	1	1	2
DS N° 609 Standards for industrial liquid residues to sewerage	Components (surface water quality)	1	1	2
DS N° 4,363 Forest law approval	Components (native plants associated to wellsprings)	2	1	3
DS N° 2,374 Standards for forest exploitation in forest watersheds and dams	Components (forests in watersheds)	1	1	2
Decree N° 878 Prohibition to extractive native fish in inland waters	Components (ichthyofauna)	1	1	2
NCh N° 1,333/87 Chilean standard for water quality	Components (water quality for aquatic life)	2	1	3
Secondary standards for environmental quality	Components (water quality for aquatic life)	3	1	4
Resolution N° 425 Groundwater exploration and exploitation standards	Components (water quality for aquatic life)	2	1	3
Resolution N° 197 Loa River reduction to new exploitation feeding	Components (water)	2	1	3
Total protection value				144

DS Supreme Decree, *DL* Law Decree, *DFL* Decree with Force of Law, *NCh* Chilean Standard, *CITES* Convention on International Trade in Endangered Species of Wild Fauna and Flora.

Möller and Muñoz-Pedreros

Möller and Muñoz-Pedreros *Revista Chilena de Historia Natural* 2014 87:23, doi:10.1186/s40693-014-0023-1

Table 2 presents legal protection value provided by legal standards to different types of inland wetlands in Chile. Zero values indicate that standard does not apply. From 47 standards identified, 46 of them apply to rivers, providing total protection value of 141; 27 apply to brackish Andean lakes, providing total protection value of 80; 44 apply to lagoons, providing total protection of 134; 28 apply to peatlands ('vegas' and 'bofedales') with total protection of 83; 32 apply to swamp forests, providing total protection of 98; and 34 apply to marshes ('bañados' and 'ciénagas'), giving full protection of 102. As shown in Table 2, greater legal protection is provided in descending order: to rivers, lagoons, marshes, swamp forests, and finally, peatlands and brackish Andean lakes have the lowest legal protection values.

Table 2: Protection value and number of legal standards to different types of inland wetlands in Chile

Legal standards/ protection value	VP potencial wetlands	Rivers	Brackish Andean lakes	Lagoons	High-altitude Andean peatlands	Freshwater swamp forests	Permanent freshwater marshes and swamps
Act N° 19,300	6	6	6	6	6	6	6
DS N° 531 Washington Convention	2	2	2	2	2	2	2
DS N° 141 CITES	2	2	2	2	2	2	2
DL N° 3,485 Ramsar Convention	5	5	5	5	5	5	5
DS N° 868 Migratory species (Bonn Convention)	2	2	2	2	2	0	2
Decree N° 1,963 Convention on Biodiversity	4	4	4	4	4	4	4
Act N° 11,402 Barriers and river banks and streams	5	5	0	0	0	0	0

Act N° 17,288 National Monuments	4	4	4	4	4	4	4
Act N° 18,892 Aquaculture and Fishing Law	5	5	0	5	0	0	0
Act N° 18,902 Sanitary health superintendency creation; industrial liquid residues control	4	4	4	4	4	4	4
Act N° 19,473 Hunting Law	4	4	4	4	4	4	4
DFL N° 208 Provisions to fishing development	3	3	0	3	0	0	0
DFL N° 701 About forest development	3	3	0	3	0	3	3
Act N° 20,283 Native Forest and Forest Development Recovering Law	6	6	0	6	0	6	6

Act N° 20,256 Recreative fishing standards	6	6	0	6	0	0	0
Act N° 20,411 Prevent constitution water rights (Arica/ Parinacota a O'Higgins)	4	4	4	4	4	4	4
DFL N° 725 Sanitary Code Art. 73	3	3	0	3	0	0	0
DFL N° 1,122 Water Code	4	4	4	4	4	4	4
DS N° 1 Aquatic pollution control rules	2	2	2	2	0	0	0
DS N° 5 Hunting law standard approval	2	2	2	2	2	2	2
DS N° 29 Approval of classification of species standards based on conservation conditions	2	2	2	2	2	2	2

DS N° 30 Act 19,300 Standards	4	4	4	4	4	4	4
DS N° 46 Groundwater discharge standards	2	2	2	2	2	2	2
DS N° 33 Classification process of species according to conservation condition	2	2	0	0	0	0	0
DS N° 41 Classification process of species according to conservation condition	2	2	0	2	2	2	2
DS N° 42 Classification process of species according to conservation condition	2	2	0	2	0.	2	2

DS N° 51 Classification process of species according to conservation condition	2	2	2	2	2	0	2
DS N° 82 Act No° 20,283 Soil, water, and wetland regulations	4	4	4	4	4	4	4
DS N° 90 Standards to liquid residue discharge into surface fresh water	2	2	2	2	0	0	0
DS N° 93 (1995) Primary and secondary environmental quality standards	3	3	3	3	3	3	3
DS N° 93 (2009) Native Forest Recovery and Forestry Development rules	4	4	0	4	0	4	4

DS N° 193 DL N° 701 General rules	2	2	0	2	0	2	2
DS N° 210 Regulation of reforestation and planting for recreational fishing	2	2	0	2	0	0	0
DS N° 236 Regulation not discharged sewage disposal to sewer networks	2	2	2	2	2	2	2
DS N° 238 Regulation for parks and marine reserves (including inland waters)	2	2	2	2	2	2	2
DS N° 320 Environmental aquaculture rules	2	2	0	2	0	0	0
DS N° 430 Consolidated, coordinated, and systematized document on fishery law	3	3	0	3	0	0	0

DS N° 594 Rules for health and environmental conditions in work places	2	2	2	2	2	2	2
DS N° 351 Rules for neutralization and depuration of liquid industrial residues	2	2	2	2	2	2	2
DS N° 609 Standard for sewerage industrial liquid residues discharge	2	2	0	2	2	2	2
DS N° 4,363 Approves final text forestry law	3	3	0	3	0	3	3
DS N° 2,374 Regulation to forest exploitation in forest watersheds and dams	2	2	0	2	0	2	2

Decree N° 878 Native fish extraction prohibition in inland waters	2	2	2	2	2	2	2
NCh N° 1,333/87 Chilean water quality standards	3	3	0	3	3	3	3
Secondary environmental quality standards	4	4	0	4	0	0	0
Resolution N° 425 Establish groundwater exploration and exploitation standards	3	3	3	3	3	3	3
Resolution N° 197 Declares Loa River reduction and feeders for new farms	3	0	3	0	3	0	0
Total standards	47	46	27	44	28	32	34
Total legal protection value	144	141	80	134	83	98	102

DS Supreme Decree, *DL* Law Decree, *DFL* Decree with Force of Law, *NCh* Chilean Standard, *CITES* Convention on International Trade in Endangered Species of Wild Fauna and Flora.

Möller and Muñoz-Pedreros

Möller and Muñoz-Pedreros *Revista Chilena de Historia Natural* 2014 87:23, doi:10.1186/s40693-014-0023-1

Restriction Value (RV) of Legal Standards

Eight legal restriction standards considered affecting conservation, and rational use of wetlands (Table 3) was identified. Some of these standards are also included in the protection regulations since their articles incorporate both protection and restriction issues for wetland conservation. Legal restriction applicability was assessed considering what and how it affects the wetland system. In this case, unlike the rules of protection, three applicability levels were not identified, since every effect caused by standards are supposed to affect the entire system or biological diversity, which was also considered at ecosystem level, so all were assigned with value 3. Table 3 also shows restriction value to conservation or sustainable use that provides laws to wetlands, and Table 4shows their application to different wetland types considered in this study. Zero values indicate that standard does not apply.

Table 3: Legal restriction standards which affects the rational use of inland wetlands in Chile, applicability (A) in terms of how and what affect to wetlands, hierarchy (J) and restriction value (RV) of each legal standard and total restriction value

Legal standard	Standard objective	How affect	What affect	A	J	RV

Act N° 18,248 Mining Law	Property acquisition of a mining concession enables to request the mining right's constitution and water right ownership. Written permission by governor is required for mining operations at lower distance than 50 yards in flood barriers, waterways, and lakes for public use.	Intervention to wetland systems, which may completely affect structure	Every type of wetlands (rivers) and lakes	3	3	6
Act N° 20,256 Standards for recreational fishing	Development of recreational fishing activity, and the economic and associated tourist activities.	Introduction of alien species affects biodiversity and system as a whole	Rivers and lakes	3	3	6
Act N° 11,402 Barriers and standardization of banks and river flows	Barrier works as well as standardization of banks and beds of rivers, lagoons and tidelands carried out with legal involvement can only be executed and planned by sanitary authorities of Public Works Ministry.	Regularization of channels and gravel mining extraction affecting the aquatic system structure	Rivers, lakes, and streams	3	3	6
	Extraction of gravel and sand from beds of rivers and streams should be carried out with municipality permits, which may charge duties or subsidies.					

Act N° 18,450 Rules for private irrigation and drainage development	Cost analysis benefits, construction, and rehabilitation of irrigation as well as drainage works to enable agricultural soils with poor drainage.	Drainage of wetlands results in complete destruction system	Swamp forests, marshes, and swamps	3	3	6
DFL N° 235	Incentive system to degraded soil recovering, proposing regulation of channels as management practices.	Regulation of channels affects the aquatic system structure	Water flows (rivers)	3	2	5
DFL N° 701	Afforestation development to soil recovery, making economic exploitation available.	Forest development results in swamp forest destruction (replacement by exotic species)	Swamp forests	3	2	5
Decree N° 193 General standards DFL N° 701	Assess lands suitable to forestry for quality forest development benefits, those corresponding to Ñadis soils.	Development of forest Ñadis soils results in destruction, affecting swamp forest (replacement by exotic species)	Swamp forests	3	1	4
Decree N° 98 Act N° 18,450	Defined as drainage of structures, elements, and tasks aimed to evacuate soil waters excess on surface or subsurface with restriction factor to culture development.	Wetland draining results in complete destruction of system	Swamp forests	3	1	4
TOTAL RESTRICTION VALUE						42

DFL Decree with Force of Law.

Möller and Muñoz-Pedreros

Möller and Muñoz-Pedreros *Revista Chilena de Historia Natural* 2014 87:23, doi:10.1186/s40693-014-0023-1

Table 4: Restriction value and number of legal standards to different inland wetlands in Chile

Legal standards	Wetland potential	Rivers	Brackish Andean lakes	Lagoons	High-altitude Andean peatlands	Freshwater swamp forests	Permanent freshwater marshes and swamps
Act N° 18,248 Mining Code	6	6	6	6	6	6	6
Act N° 20,256 Recreational fishing standards	6	6	0	6	0	0	0
Act N°11,402 Barriers and works for bank and channel standardization	6	6	0	6	0	0	0
Act N° 18,450 Standards for private investment in watering and drainage systems	6	0	0	0	0	6	6
DFL N° 235	5	5	0	0	0	0	0
DFL N 701 Forest development	5	0	0	0	0	5	0

Decree N° 193 General rules DL N° 701 About forest development	4	0	0	0	0	4	0
Decree N° 98 Act N° 18,450 Development of private investment in watering and drainage systems	4	0	0	0	0	4	0
Total number of standards	8	5	1	3	1	5	2
Total value of legal restrictions	42	28	6	18	6	25	12

DL Decree Law, *DFL* Decree with Force of Law.

Möller and Muñoz-Pedreros

Möller and Muñoz-Pedreros *Revista Chilena de Historia Natural* 2014 87:23, doi:10.1186/s40693-014-0023-1

Restriction to sustainable use of wetlands established by legal standards to different types of inland wetlands reveals that the eight identified standards provide a total restriction value of 42. Wetlands mostly affected are rivers and swamp forests, both with five standards, and restriction values of 28 and 25, respectively, followed by lagoons with three rules and restriction value of 18 and marshes with two rules and a restriction value of 12. Least affected are brackish Andean lakes and peatlands to which a single standard applies, giving restriction value 6.

Real Protection Value (RPV)

Real protection value (RPV) was estimated once the number of legal rules applied to each type of wetland is established (Figure 1). This value is obtained by summing protection values (PVs), given by all legal standards applied to each type of wetland; and subtracting restriction values (RV), given by all legal standards that apply to the same wetland, as follows: $RPV = \Sigma PV - \Sigma RV$. Distribution of values is grouped into three ranges, giving following nominal values: value 1 to less protected wetlands, value 2 to intermediate protected ones, and value 3 to the best legally protected wetlands. The real protection value (RPV) given by the set of laws applied to different types of wetlands in the study is presented in Table 5. This (RPV) is deployed between 73 and 116. These values were grouped into ranges to where numeric values between 1 and 3 (Table 6) were assigned, being 3 the highest protection value and 1 the lowest protection value.

Table 5: Nominal and numerical real protection granted by legal rules to different types of inland wetlands in Chile

Type of wetland	Total protection value	Protection value	Total restriction standards	Restriction value	Real protection value	Numeric protection value
Wetlands	47	144	8	42	102	3
Rivers	46	141	5	28	113	3
Brackish Andean lakes	27	80	1	6	74	1
Lagoons	44	134	3	18	116	3

High-altitude Andean peatlands	28	83	1	6	77	1
Swamp forests	32	98	5	25	73	1
Permanent freshwater marshes and swamps	34	102	2	12	90	2

Möller and Muñoz-Pedreros

Möller and Muñoz-Pedreros *Revista Chilena de Historia Natural* 2014 87:23, doi:10.1186/s40693-014-0023-1

Table 6: Nominal and numerical protection value granted by law to different types of wetlands

Nominal value	Rate	Real protection number value	Type of wetland
Low protection	≤87	1	Brackish Andean lakes
			High-altitude Andean peatlands
			Swamp forests
Intermediate protection	88–102	2	Permanent freshwater marshes and swamps
High protection	≥103	3	Rivers and lagoons

Möller and Muñoz-Pedreros

Möller and Muñoz-Pedreros *Revista Chilena de Historia Natural* 2014 87:23, doi:10.1186/s40693-014-0023-1

Real protection provided by laws is higher (value 3) in rivers and lagoons, intermediate (value 2) in marshes, and low (value 1) in brackish Andean lakes, peatlands, and swamp forest. So, the best protected wetlands, according to the number and legal hierarchy affecting direct or indirect protection, are rivers and lagoons. Wetlands with intermediate protection category are marshes and peatlands. Finally, the least protected are brackish Andean lakes, peatlands, and swamp forests.

DISCUSSION

General review of legal standards affecting conservation of wetlands in Chile reveals that before year 2012, there were no laws addressed to wetlands as ecosystem, since they only involved some components (water, flora, wildlife). The first reference to wetland concept in the Chilean law, except acts related to Ramsar Convention adhesion enacted in 1981, appears in 1994 with Act No. 19,300, one time mentioned.

Same way, wetland definition arises for the first time in 2011 by act of Supreme Decree N 82, Soil, water, and Wetlands Regulations Act No. 20,283. This is the evidence of the lack of consideration and the late inclusion of these ecosystems in the country regulations.

There is a different legal protection to different types of inland wetlands since current legislation in Chile does not provide equal protection to the different types of wetlands. Lagoons and rivers have the highest legal protection values due to the large number of regulations applied favorably to these systems and fewer standards that affect them adversely. For marshes, the intermediate value of legal protection comes from that; in spite, there are a number of regulations favoring conservation, and there are a few restrictions affecting them.

Andean systems (brackish lakes and peatlands) present different condition; since in spite a few restriction standards affecting them, they are poorly protected by current legislation. Swamp forest is the most dramatic case; which in spite of important existing regulatory standards that should protect them, they are the most affected ones by the restriction rules. The greatest impacts are Act No. 18,450 which promotes irrigation and drainage and Decree Law No. 701 about forest development. These regulations subsidize up to 75% of activities which enable the poorly drainaged lands to become agricultural territory, which means the main threat to swamp forests of coastal watersheds in La Araucanía region (Urrutia [2005], García [2005]) and its evolution towards greater fragmentation in recent years (Peña-Cortés et al. [2011]).

Legal standards above mentioned are clear perverse incentives and disincentives to the rational use of wetlands in Chile.

One aspect considered relevant in the Ramsar Convention was to assess effectiveness of legislative and institutional measures related to promote the conservation and the rational use of wetlands, are sectoral legal and institutional measures affecting wetlands such as financial and tax incentives to convert them. Nelson ([1986]) noted that wetland policies in the US and UK had administrative divisions with antagonistic competences. So, while the US Department of Agriculture provided incentives to drainage, the Interior Department was promoting conservation of wetlands, same way that the Ministry of Agriculture, Forestry, and Fisheries and the Nature Preservation Council faced each other in the UK. This problem was addressed in the US by removing some incentives for wetland drainage, which also included other conservation and education measures (Dahl [2000]), reducing by 80% the wetland loss tendency between 1986 and 1997 compared to previous decade. Apparently, some Latin American countries have also advanced on this subject. For example in Ecuador, the legislation review by Echeverría ([2008]) about wetland management does not identify measures that indirectly support the loss or degradation of wetlands through negative incentives.

Review of some national reports presented by countries as contracting party in the last conference of the Ramsar Convention, held in Bucharest, Romania in 2012[a] (COP11), allowed to know measures that countries reported as implemented regarding incentives which promote the conservation and the rational use of wetlands, as well as those intended to remove perverse incentives which discourage them.

[a]http://www.ramsar.org/library/field_documents%253Afield_language/english-1/field_tag_body_event/cop11-bucharest-2012-415

USA is the American country with the highest amount of initiatives on this subject. Report to COP11[b] indicates several provisions that discourage conversion of wetlands to cropland. The federal Swampbuster policy, provision of Food Security Law dated 1985, has eliminated the policy incentives and other mechanisms which have technically and economically allowed the destruction of wetlands, acting and separating those who make it from other agricultural policy benefits. Federal efforts to restore wetlands have increased from 1987, with legal standards of critical preservation and restoration with two programs which stopped and even reversed loss as Wetland Conservation (WC) provisions of Agricultural Law in 1985 and the Wetlands Reserve Program (WRP) of Agricultural Law in 1990 (Coperland [2010]).

[b]http://www.ramsar.org/sites/default/files/documents/pdf/cop11/nr/cop11-nr-usa.pdf

In national report, Canada[c] indicates that measures to eliminate perverse incentives in some provinces consist of wetlands filling prevention rules which enable marginal agricultural soils and balance benefits provided by wetland drain restoration projects with cleaning, storage, and discharge of water functions fulfilled by wetlands. The UK[d] works reforms to agricultural policy to have a positive impact on marsh wetlands used for grazing by agri-environment patterns, management agreements, and other keeping administration systems.

[c]http://www.ramsar.org/sites/default/files/documents/pdf/cop11/nr/cop11-nr-canada-e.pdf

[d]http://www.ramsar.org/sites/default/files/documents/pdf/cop11/nr/cop11-nr-uk.pdf

The China[e] report indicates that diverse level governments and their departments have a strict control over every kind of activity that damages wetlands by enacting 11 provincial (autonomous regions) wetland conservation regulations, which prohibit several activities and perverse incentives. People convocation has taken the surveillance and control in the implementation of those laws at various levels. Australia[f] has already indicated in the COP10 national report 2008 about the National Water Initiative, Water for Future, and Water Law 2007 developed to stop perverse incentives in the water management. Same way, New Zealand[g] declared in national report that they have eliminated subsidies to land development.

[e]http://www.ramsar.org/sites/default/files/documents/pdf/cop11/nr/cop11-nr-china.pdf

[f]http://www.ramsar.org/sites/default/files/documents/pdf/cop11/nr/cop11-nr-australia.pdf

[g]http://www.ramsar.org/sites/default/files/documents/pdf/cop11/nr/cop11-nr-newzealand.pdf

In South American countries[h], Bolivia, Colombia, Peru, and Chile state, they did not take measures to promote incentives to encourage conservation and rational use of wetlands. Countries showing incentives are Costa Rica by promoting private refuges, territory tax payment exemption, and payment for environmental services; Ecuador by applying tax exemption to rural land containing wetlands and monetary incentives to private owners and community which preserve Andean wetlands; and Argentina and Venezuela by providing financial incentives for conservation and planning in land management.

[h]http://www.ramsar.org/library/field_documents%253Afield_language/espa%C3%B1ol-4/field_tag_body_event/cop11-bucharest-2012-415/field_tag_countries/neotropics-15?search_api_views_fulltext=&items_per_page=20

Regarding measures to remove perverse incentives, Bolivia, Brazil, Costa Rica, and Peru declare that they have not taken them. Ecuador and Venezuela note that they count with these measures. Uruguay, Colombia, and Chile indicate that these measures are being planned and involve payments for environmental services and penalties for damage to the ecosystem. They consider compensatory wetland restoration. Analysis of measures reveals that these have to do with the creation of new incentives rather than eliminating those perverse incentives which favor, fill, and convert wetlands to agricultural soils or urban with bonus or by other means (see national reports of countries mentioned[i]).

Based on the formerly mentioned, three actions must be implemented in Chile to improve conservation and rational use of wetlands: 1) Repeal rules which promote destruction of wetlands through incentives. 2) Discuss a special law for proper management and conservation of wetlands, including tax incentives and the corresponding regulation to operate. 3) Prioritize conservation actions for wetland types with low protection value such as swamp forests, peatlands, and brackish Andean lakes.

CONCLUSIONS

In Chile practically, there are no specific legal regulations on wetlands; there are only some decrees which applied indirectly to these ecosystems, addressed only to some components (water quality and quantity and wild species), without ecosystem approach. Current legal regulations in Chile do not allow a proper protection of wetlands and rational use as mandated by the committed country as a member of the Ramsar Convention. In Chile, to improve wetland conservation conditions, every incentive promoting wetland destruction must be eliminated; promulgate specific regulations for proper management and conservation of wetlands including tax incentives and corresponding operating regulations; and favor conservation actions to less protected wetland, such as swamp forests, peatlands, and brackish Andean lakes.

AUTHORS' CONTRIBUTIONS

PM developed the methodology, conducted legal review and drafted the manuscript. AM participated in the design of the methodology, in writing and revision of the manuscript. Both authors read and approved the final manuscript.

ACKNOWLEDGEMENTS

We thank to projects: management and restoration measures of wetland ecosystems in southern Chile, Ministry of Environment (2011-2012); identification of potential areas to establish reserves aimed to the protection of native fauna of freshwater aquatic species, FIP No. 2008-58, Fisheries Research Fund (2009-2010); analysis of biodiversity in Antofagasta Region (2007-2008) CONAMA, FNDR, all of them executed by Center for Agricultural and Environmental Studies. Also, we thank Charif Tala (Ministry of Environment) for his comments and suggestions to methodology. AMP appreciates the contribution of the General Direction for Research and Graduate Studies at the Catholic University of Temuco, DGIPUCT project No. CD2010-01 and Project 0804 MECESUP UCT.

REFERENCES

1. Abell R: Conservation biology for the biodiversity crisis: a freshwater follow-up *Conserv Biol* 2002, 16:1435-1437.
2. Abell R, Allan JD, Lehner B: Unlocking the potential of protected areas for freshwaters. *Biol Conserv* 2007, 134:48-63
3. Asenjo R: Institucionalidad pública y gestión ambiental en Chile. Expansiva, serie. *Foco* 2006, 91.

4. Bahamonde N, Carvacho A, Jara C, López M, Ponce F, Ma Retamal E, Rudolph E:Categorías de conservación de decápodos nativos de aguas continentales de Chile. *Bol Mus Nac Hist Nat* 1998, 47:91-100.

5. Barbier E, Acreman M, Knowler D: *Valoración económica de los humedales. Guía para decidores y planificadores*. Oficina de la Convención de Ramsar, Gland, Suiza; 1997.

6. Brañes R (2000) Manual de Derecho Ambiental. Fundación Mexicana para la educación Ambiental, Editado por Fondo de Cultura Económica, 2° edición. Ciudad de México

7. Bravo D: Conservación y preservación de los humedales en Chile. *Justicia ambiental: Revista de Derecho Ambiental de Fiscalía del Medio Ambiente/FIMA*2010, 2:91-158.

8. Castillo M: *Régimen jurídico de protección del medio ambiente. Aspectos generales y penales. Segunda edición. Ediciones Bloc, Santiago de Chile*. 1994.

9. *Estudio identificación de áreas potenciales para establecer reservas destinadas a proteger la fauna nativa de especies hidrobiológicas de agua dulce Centro de Estudios Agrarios y Ambientales/ Fondo de Investigación Pesquera*. 2010.

10. *Catastro y evaluación de los recursos vegetacionales nativos de Chile*. Editado por CONAF/CONAMA, Chile; 1997.

11. *Aire, flora, fauna, áreas silvestres protegidas, procedimientos administrativos ambientales*. 1996.

12. *Diversidad biológica, identificación y diagnóstico preliminar del ordenamiento jurídico aplicable a la protección de la diversidad biológica*. 1996.

13. *Áreas silvestres protegidas, legislación punitiva y sancionatoria ambiental*. 1997.

14. *Áreas silvestres protegidas, legislación sancionatoria ambiental*. 1997.

15. *Ley sobre bases generales del medio ambiente y sus reglamentos*. Chile, Editado por Comisión Nacional del Medio Ambiente, Santiago; 2009.

16. Coperland C: *Wetlands: an overview of issues. Congressional Research Service Reports. Paper 37*. 2010.

17. *Status and trends of wetlands in the conterminous United States 1986 to 1997*. U.S. Department of the Interior, Fish and Wildlife Service, Washington, D.C; 2000.
18. Díaz-Paéz E, Ortiz J: Evaluación del Estado de conservación de los anfibios en Chile. *Rev Chil Hist Nat* 2003, 76:509-525.
19. *Directrices para la aplicación de las categorías de gestión de áreas protegidas*. UICN Gland, Suiza; 2008.
20. Dugan P: *Conservación de Humedales. Un análisis de temas de actualidad y acciones necesarias*. UICN, Suiza; 1992.
21. Echeverría H: *Proyecto Capacitación en base de una Revisión de la Legislación e Institucionalidad relacionada con la Gestión de Humedales en el Ecuador (WFF/06/EC/2)*. 2008.
22. Fernández P: *Manual de Derecho Ambiental Chileno. Editorial Jurídica de Chile*. 2004.
23. Figueroa R, Palma A, Ruiz V, Niell X: Análisis comparativo de **índices** bióticos utilizados en la evaluación de la calidad de las aguas en un río mediterráneo de Chile: río Chillán, VIII Región. *Rev Chil Hist Nat* 2007, 80(2):225-242.
24. Gallardo E: Legislación sobre protección de la flora arbórea y arbustiva nativa chilena. In *Libro rojo de la flora terrestre de Chile (Primera Parte)*. Edited by Benoit IL. Corporación Nacional Forestal, Santiago de Chile; 1985.
25. García A: *Análisis estructural y estado de conservación de los rodales costeros de temu y pitra, entre Imperial y Queule, propuestas para su conservacion*. Tesis, Universidad Católica de Temuco, Chile; 2005.
26. Hermosilla J: *Tratados Internacionales vigentes en Chile en materia de protección a la Biodiversidad y su relación con la legislación interna. Facultad de Ciencias Jurídicas y Sociales*. Escuela de Derecho Universidad Austral de Chile, Valdivia, Chile; 2004.
27. Iza AO, Rovere MB: *Gobernanza del agua en América del Sur: dimensión ambiental*. UICN, Gland, Suiza y Cambridge, Reino Unido; 2006.
28. Linstone H, Turoff M: *The Delphi Method. Techniques and Applications*. Addison-Wesley, Boston, United States of America; 1975.

29. López-Lanús B, Blanco DE: *El Censo Neotropical de Aves Acuáticas 2004. Global Series No. 17*. Wetlands International, Buenos Aires, Argentina; 2005.
30. Marín V, Delgado L, Vila I: Sistemas Acuáticos, ecosistemas y cuencas hidrográficas. In *Macrófitas y vertebrados de los sistemas límnicos de Chile*. Edited by Vila I, Veloso A, Schlatter R, Ramírez C. Editorial Universitária, Santiago de Chile; 2006.
31. Mitsch WJ, Gosselink JG: *Wetlands third edition*. John Wiley, New York; 2000.
32. Muñoz-Pedreros A: Los humedales del río Cruces y la Convención de Ramsar. Un intento de protección fallido. *Gest Amb* 2004, 10:11-26.
33. Nelson RW: Wetlands policy crisis: United States and United Kingdom. *Agr Ecosyst Environ* 1986, 18:95-121.
34. Novitzki R, Smith RD, Fretwell JD: Wetland Functions, Values and Assessment. In*(comp). National water summary on wetland resources. United States Geological Survey*. Edited by Fretwell JD, Williams JS, Redman PJ. Water Supply Paper 2425, Washington, D.C; 1996.
35. Olivares A: El nuevo marco institucional ambiental en Chile. *Revi Cat Dret Amb* 2010, 1(1):1-23.
36. Ortiz S: *Legislación vigente en el ámbito del manejo de cuencas hidrográficas*. Corporación Nacional Forestal Octava Región, Concepción, Chile; 1986.
37. Ortiz S: *La actividad forestal y su regulación ambiental*. Seminario CORMA sobre Legislación forestal y su dimensión en el derecho ambiental, Concepción, Chile; 1990.
38. Peña-Cortés FA, Gutierrez PL, Rebolledo GD, Escalona MA, Bertran CE, Schlatter R, Tapia J:Determinación del nivel de antropización de humedales como criterio para la planificación ecológica de la cuenca del lago Budi, Chile. *Rev Geog Nor Gr* 2006, 36:75-91.
39. Peña-Cortés FA, Pincheira-Ulbrich J, Escalona-Ulloa M, Rebolledo G: Cambio de uso del suelo en los geosistemas de la cuenca costera del río Boroa (Chile) entre 1994 y 2004. *Rev Fac Cienc Agr Univ Nac Cuyo* 2011, 43(2):1-20.

40. Pérez-Losada M, Jara CG, Bond-Buckup G, Crandall KA: Conservation phylogenetics of Chilean freshwater crabs *Aegla* (Anomura, Aeglidae): assigning priorities for aquatic habitat protection. *Biol Conserv* 2002, 105:345-353. Publisher Full Text

41. Ramírez C, San Martín C, Rubilar H: Una propuesta para la clasificación de los humedales chilenos.\ *Rev Geog Valp* 2002, 33:265-273.

42. *Marcos para manejar humedales de importancia internacional y otros humedales. Con comprendidos de los Lineamientos adoptados por la Conferencia de las Partes Contratantes en sus reuniones 4a.,5a.,6a., y 7a*. Oficina de la Convención de Ramsar, Gland, Suiza; 2000.

43. Rodríguez-Becerra M, Espinoza G: Gestión ambiental en América Latina y el Caribe: evolución, tendencias y principales prácticas. In *Inter Division de Medio Ambiente, Banco Interamericano de Desarrollo* Edited by David W. 2002.

44. *Cuadernos de Análisis Jurídico, Colección Derecho Ambiental I. Programa de Derecho y Política Ambiental. Facultad de Derecho.* Universidad Diego Portales, Santiago de Chile; 2011.

45. Schlatter RP, Espinosa LA, Vilina Y: Coasts of central and southern Chile (Region 15).Chapter 3.3.3. Los Humedales de América del Sur, una agenda para la conservación de su Biodiversidad y las Políticas de Desarrollo.\ In *Resumen Ejecutivo. Wetlands International: 14 y CD. Book* Edited by Canevari P, Davidson I, Blanco DE, Castro G, Bucher EH. 2001.

46. *Inventario de Humedales de la Región Neotropical*. IWRB y UICN, Reino Unido; 1986.

47. Urrutia O: *Estado de conservación de los bosques pantanosos y su relación con los ecosistemas asociados en el borde costero, entre Imperial y Queule, IX Región*. Tesis, Universidad Católica de Temuco, Chile; 2005.

48. Valdovinos C: Moluscos terrestres y dulceacuícolas de la cordillera de La Costa chilena. In *Historia, biodiversidad y ecología de los bosques costeros de Chile*. Edited by Smith-Ramírez C, Armesto JJ, Valdovinos C. Editorial Universitaria, Santiago de Chile; 2005.

49. Valenzuela R: *Diagnóstico preliminar del ordenamiento jurídico vigente, en lo que se refiere a la protección de la diversidad biológica*. 1994.

50. Veloso A: Batracios de las cuencas hidrográficas de Chile: origen, diversidad y estado de conservación. In *Macrófitas y vertebrados de los sistemas límnicos de Chile*. Edited by Vila I, Veloso A, Schlatter R, Ramírez C. Editorial Universitaria, Santiago de Chile; 2006.
51. Vila I, Pardo R, Dyer B, Habit E: Peces límnicos, diversidad, origen y estado de conservación. In *Macrófitas y vertebrados de los sistemas límnicos de Chile*. Edited by Vila I, Veloso A, Schlatter R, Ramírez C. Editorial Universitaria, Santiago de Chile; 2006.
52. Woodward RT, Wui YS: The economic value of wetland services: a meta-analysis. *Ecol Econ* 2001, 37:257-270.
53. Zegers G, Larraín J, Díaz F, Armesto J: Impacto ecológico y social, de la explotación de pomponales y turberas de *Sphagnum* en la isla Grande de Chiloé. *Rev Amb Des* 2006, 22:28-34.

Chapter 4

Using Strategically Applied Grazing to Manage Invasive Alien Plants in Novel Grasslands

Jennifer Firn[1], Jodi N Price[2], and Ralph DB Whalley[3]

[1]School of Earth, Environmental and Biological Sciences, Queensland University of Technology, Brisbane, QLD 4001, Australia

[2]School of Plant Biology, University of Western Australia, Crawley, WA 6009, Australia

[3]Botany, University of New England, Armidale, NSW 2352, Australia

ABSTRACT

Introduction

Novel ecosystems that contain new combinations of invasive alien plants (IAPs) present a challenge for managers. Yet, control strategies that focus on the removal of the invasive species and/or restoring historical disturbance regimes often do not provide the best outcome for long-term control of IAPs and the promotion of more desirable plant species.

Methods

This study seeks to identify the primary drivers of grassland invasion to then inform management practices toward the restoration of native ecosystems. By revisiting both published and unpublished data from experiments and case studies within mainly an Australian context for native grassland management, we show how alternative states models can help to design control strategies to manage undesirable IAPs by manipulating grazing pressure.

Results

Ungulate grazing is generally considered antithetical to invasive species management because in many countries where livestock production is a relatively new disturbance to grasslands (such as in Australia and New Zealand as well as Canada and the USA), selective grazing pressure may have facilitated opportunities for IAPs to establish. We find that grazing stock can be used to manipulate species composition in favour of the desirable components in pastures, but whether grazing is rested or strategically applied depends on the management goal, sizes of populations of the IAP and more desirable species, and climatic and edaphic conditions.

Conclusions

Based on our findings, we integrated these relationships to develop a testable framework for managing IAPs with strategic grazing that considers both the current state of the plant community and the desired future state—i.e. the application of the principles behind reclamation, rehabilitation, restoration or all three—over time.

REVIEW

Introduction

Research into invasive alien plants (IAPs) has largely been focused on understanding *why* and *how*some plant species can colonise, persist, expand their ranges, and sometimes even dominate plant communities over relatively short evolutionary timeframes—this, often under the guise of informing and improving the design of control strategies (Sax and Brown 2000; Levine et al. 2003; MacDougall and Turkington 2005; Vila et al. 2011). Yet, the most common control strategies remain centred on removing IAPs with various rudimentary physical, chemical or biological measures, such as bulldozers, herbicides and insects. In some control programs, particularly within an Australian context for land management, strategies are put into place to return historical disturbance regimes, e.g. by excluding grazing and/or prescribing seasonal burning. Despite these efforts, many control programs continue to prove ineffectual, resulting in the re-establishment of the same IAP or others (Gardener et al. 2010), a phenomenon referred to as *filling a 'weed-shaped' hole* (Firn et al. 2008). That being said, common control strategies—e.g. where the IAP is killed or removed and the historical/bioregional disturbance regime applied—are largely based on traditional models of succession. These models suggest that plant assemblages are dynamic but still progress towards a final state (or climax community) along a continuum that is regulated by internal forces (i.e. changes resulting from interactions among organisms and the physical environment) as well as climatic conditions. Disturbance and/or intra- and inter-specific competition can shift vegetation community composition forward or backward along a continuum (Clements 1916). This notion of a continuum of states suggests that alterations to a plant community are reversible, yet we now know that this is not always the case (cf. Perring et al.2013). In addition, weed control has commonly been approached from principles of population ecology, being focused on the behaviour and requirements of single species, as opposed to the holistic behaviour of vegetation communities undertaking active recruitment following control measures—the so-called and aforementioned *weed-shaped hole* (Booth et al. 2003).

The problem with these approaches is that the original community is usually not restored once the causal disturbance has been reduced or mitigated; instead, what mostly results is a new assemblage of plant species, not predicted along the historical continuum (cf. Suding et al. 2004). Novel ecosystems and the underlying theory (representing a central theme for this special issue) have recently (and contentiously) arisen in the field of restoration ecology to address the unique management issues that are presented by these altered systems (Hobbs et al. 2006, 2009). As discussed further below, this is particularly relevant to IAP-dominated ecosystems in which many new assemblages are achieved by human-mediated dispersal (Richardson and Gaertner 2013).

Novel Ecosystems and Alternative States Models

Recent novel ecosystem theory proposes that threshold models such as alternative states models are effective restoration tools for managing systems altered by direct and indirect human-induced disturbances (Hobbs and Suding 2009), particularly since not all novel ecosystems are necessarily desirable (Richardson and Gaertner 2013). In this case, the goal may be, e.g., to alter the conditions of the novel ecosystem to a more desirable state. Alternative states models suggest that historical disturbance regimes will not work where the functioning of the ecosystem has changed (i.e. nutrient cycling, hydrology and energy flows; Zavaleta et al. 2001; Suding et al.2004; Hobbs et al. 2006). These models have evolved within the context of rangeland ecology (Westoby et al. 1989; Cingolani et al. 2005), restoration ecology (Suding et al. 2004) and ecological resilience theory (Gunderson 2000) but have only recently been applied within invasion ecology (Cox and Allen 2008; Firn et al. 2010; Gaertner et al. 2012). In contrast to traditional models of succession, alternative states models explain how different communities can exist at the same site depending on conditions. Transitions between states are regulated by both internal forces (that are also included in succession models) as well as external ones. External forces are defined as newly imposed or alterations to existing disturbance(s) and climatic conditions. For instance, if a community has transitioned to a new state and the feedback relationships have shifted permanently from negative to positive within this state, restoring it back to the previous state may no longer be possible simply by reducing or mitigating the causative disturbance (Richardson and Gaertner 2013). Instead, alternative management actions may be required to initiate a transition, and the disturbance applied to initiate a transition may have little to do with the historical disturbance regime (Suding et al. 2004). Understanding the historical context of *why* and *how* IAPs dominate is important for the design of preventative measures, but this is only one piece of the puzzle. To use alternative states models as the underlying framework for invasive species management bringing together fields of restoration and invasion ecology (Hobbs and Richardson2011), we then need:

- An understanding of how the new IAP-dominated ecosystem functions (Hobbs et al. 2006; Richardson and Gaertner 2013), and
- A clear, realistic and flexible set of restoration goals aimed at gradually shifting the community towards a more desirable state over the short and long term (Hobbs 2007; Hobbs and Suding2009).

This is particularly important when IAPs are considered transformers (or drivers of ecological change) because they may alter both subtle and obvious components of ecosystem functions and services, e.g. by being excessive users of nutrients, donors of limiting nutrients, fire promoters/suppressors or salt accumulators/redistributors (Richardson et al. 2000; Richardson and Gaertner 2013). Otherwise investing time, effort and money into the application of common control strategies may lead to even further degradation (Zavaleta et al. 2001).

Methods

Objectives, Hypotheses and Investigative Approach

Human activities such as recently introduced ungulate grazing imposed at high rates and/or intensities in either productive or unproductive regions (or, likewise, regions prone to drought—as in many parts of Australia) may lead to the dominance of undesirable plant species

(Cingolani et al. 2005). These species, which are typically difficult to control over expansive landscapes (e.g.*Nitraria billardieri* in semi-arid NSW; Noble and Whalley 1978), are generally either unpalatable or respond quickly to disturbance through vegetative growth and/or germination. Even though many restoration efforts stipulate biodiversity conservation as a main objective, a solution to this management dilemma is not necessarily to remove grazing entirely (as in traditional models of succession); instead, some evidence suggests that modifying grazing pressure either (a) directly through the manipulation of the timing, frequency and intensity of grazing or (b) indirectly through the manipulation of resource availability to increase the palatability of the dominant IAP are more effective solutions (or alternative states) at least in the short term (Davies et al. 2009; Firn et al.2010). We know from extensive research that, on the one hand, grazing can reduce species richness and abundance where herbivores are exotic or newly introduced to an area (particularly in regions that are low in nutrients and rainfall) and where grazing pressure is high in both intensity and frequency (Milchunas et al. 1988; Cingolani et al. 2005). On the other hand, grazing can increase species diversity where herbivores are native to an area (particularly in regions that have moderate to high nutrients and rainfall) and where grazing pressure is low in both intensity and frequency (Milchunas et al. 1988; Olff and Ritchie 1998; Cingolani et al. 2005; Parker et al.2006; Hillebrand et al. 2007). Grazing can also produce genetic changes within native grass species, for instance, where there are marked spatial differences in grazing pressure or in the distribution of introduced perennial pasture grasses within a paddock (Scott and Whalley 1984; Magcale-Macandog and Whalley 2000, 2007). This is particularly true of Australian native grasses because of the breeding systems that many of them have evolved over geological time (Whalley et al. 2011) Based on this evidence, it is suggested that grazing, if used strategically, could then facilitate increased native diversity (Lunt et al. 2007).

In this study, we synthesize evidence that strategically applied grazing can be used successfully for the control of invasive plants and the return of more desirable species, in particular natives. By revisiting both published and unpublished data from studies of invasive species within an Australian context for grassland management, we describe two management options: (1) directly manipulating grazing pressure and (2) indirectly manipulating grazing pressure. We then make recommendations on scenarios where direct and indirect manipulation of grazing pressure can be applied. Finally, we suggest a set of generic management goals depending on the current and future desirable state of the grassland or pasture, in accordance with the principles of adaptive management. Studies included in our retrospective analysis were selected if they included a treatment where grazing was manipulated to control an IAP. This provided a suitable investigative context to ultimately present our testable conceptual depiction of alternative states models work as a theoretical underpinning for IAP control using grazing management. Detailed methodology explaining data collection and analyses for all of the presented data.

RESULTS AND DISCUSSION

Control Strategies

Directly Manipulating Grazing Pressure

To reduce IAP dominance with strategic grazing, management needs to alter those variables that enable species persistence under grazing, i.e. selectivity and palatability. There are several ways in which selectivity and/or palatability can be altered by directly modifying grazing pressure. Grazing intensity can be increased to reduce selectivity and increase the uniformity of grazing among the desirable and undesirable species. The resulting community composition will depend on the ability of the species to recover from grazing in the rest period following defoliation (Gardener et al. 2005). High intensity rotational grazing has been used successfully for the control of several invasive species (e.g. *Cirsium arvense*; De Bruijn and Bork 2006) and *Chrysanthemum leucanthemum* (Olson et al. 1997)). Pywell et al. (2010) found the most effective control strategy for *Cirsium arvense* was low intensity grazing because it maintained a competitive cover of other plant species. Many unpalatable species have life-history stages that are palatable and sensitive to grazing, and the timing of grazing (and the intensity) can be altered to periods when the invasive species is most susceptible (Hartley et al. 1984; Thomsen et al. 1993; Rinella and Hileman 2009). The duration of grazing is also important as many invasive species require repeated defoliations to have an effect (e.g. leafy spurge; Kirby et al. 1997). Grazing has been used strategically for invasive species control in two ways: (1) biocontrol, where grazing animals are used to reduce the growth, survival or reproduction of invasive plants (Popay and Field 1996), and (2) competition, by increasing the cover of resident species to suppress invasive species.

An Example of When Grazing Management can Aid in Invasive Species Management is with Thistle

Thistle species are generally unpalatable and as such tend to increase in grazed systems. However, strategic grazing can reduce thistle densities by targeting the life-history stage when the shoots are palatable (newly emerging) and beyond the rosette stage (available to stock) (Grace et al. 2002; Holst et al. 2004; De Bruijn and Bork 2006; Wallace et al. 2008). The timing of grazing that provides the best control will depend on the species as well as environmental variables. Grazing intensity and duration are important; low intensity continuous grazing increased Canadian thistle populations by favouring selective grazing, whereas high intensity rotational grazing reduced populations (De Bruijn and Bork 2006). High intensity rotational grazing also minimised the impact on resident species, which increased competition. Rest from grazing in autumn to increase competition altered the morphology of *Carduus* spp. causing plants to become more upright and accessible to stock, and grazing at this stage provided control (Bendall 1973). Control methods that indirectly alter grazing pressure have also provided thistle control, e.g. applying herbicides can alter physiology, vigour and palatability (Holst et al. 2004; Shea et al.2006). Thistles are a well studied example of an invasive species which increases under grazing but can be controlled with modifications to the grazing regime.

Resting from grazing can also be beneficial for the development of competitive species that may be more palatable than the target weed (Lodge and Whalley 1985. A number of studies have found rest from grazing can reduce the abundance of undesirable species and increase the abundance of desirable species by matching rest periods with species phenologies (e.g. Ash et al. 2011; Kemp et al. 1996). This strategy will depend, however, on whether desirable species remain in the community, their phenology (Kemp et al. 1996; Fischer et al. 2009) and whether the dominance of the IAP has been reduced prior to resting (Lunt et al. 2007). Rest from grazing in novel ecosystems also may not work because remaining original species may have experienced genetic changes that allow them to survive and persist with grazing pressure (Scott and Whalley 1984; Magcale-Macandog and Whalley 2000, 2007; Waters et al. 2010, 2011), but they may then lack the genetic variability to persist without grazing. Original species that are lost or have become rare and endangered within the ecosystem may have had inflexible breeding systems that meant they were incapable of responding to grazing pressure (Yu et al. 2000, 2003). Several studies have found that excluding grazing from sites where populations of an unpalatable plant species are high indeed favour the further expansion of the IAP (Firn et al. 2010; Hayes and Holl 2011; Isbell and Wilsey 2011; Price et al. 2011b). In these cases, it may be that the undesirable species itself should be removed using herbicides, the palatability of the species be modified by using fertilisers or some other form of disturbance occur to initially perturb the population and open opportunities for other species.

Indirectly Manipulating Grazing Pressure

Selectivity and palatability can also be altered by indirectly modifying grazing pressure. Prescribed burning within IAP-dominated ecosystems will break down mature tissue and litter, which in turn will increase light and nutrient availability and stimulate germination from the seed bank and vegetative growth. Because of this flush of resources, plant species that colonise after fire will be more palatable and nutrient rich, and these patches will then attract grazing stock and reduce selective pressure (Cummings et al. 2007). In some cases, prescribed burning may reduce dominant IAPs and promote other desirable species without livestock grazing being necessary, although evidence has shown that prescribed burning is more effective when incorporated into an integrated vegetation management program (DiTomaso et al. 2006).

Lespedeza cuneata is a perennial legume that was introduced into the USA as a forage species specifically because the species has persistence mechanisms that maintain its dominance under grazing.

Patch burning was found to reduce invasion in tallgrass prairies by maintaining the invasive species in a young and palatable stage of growth (Cummings et al. 2007). The size of the patch that is burnt matters, however, with small burnt patches more likely than large ones to result in increased grazing pressure on all species present not just the most palatable. Applying herbicides or fertilisers can also alter the palatability of IAPs. Fertilisers have been used to increase the palatability of *Eragrostis curvula*, and in combination with grazing management provided good control (Firn et al. 2010). A treatment combination of herbicides and grazing has been effective for the control of *Echium plantaginuem* (Huwer et al. 2005) and *Carduus nutans*(Shea et al. 2006). Herbicides alter the morphology of *E. plantaginuem* making the rosettes more erect and available to grazing stock (Piggin 1979).

Alternative States Models Work as a Theoretical Underpinning for IAP Control Using Grazing Management

Most of the literature on grazing management and invasive species control has been conducted in rangelands or pastures where conservation of biodiversity or native dominance is secondary to production and where the major land use is grazing. Because of this, in many cases grazing exclusion has not been experimentally tested as a management option, rather just the grazing regime has been modified. This means that models that incorporate restoring historical conditions are largely untested in the bulk of the agricultural literature. What is evident, however, from this literature is that sometimes a modification of the current grazing regime may be what is needed to provide control of invasive species. In addition to this, some recent studies have found that restoring historical conditions has not provided the best control options, supporting the notion that novel ecosystems may need novel management options (Seastedt et al. 2008).

Restoration of historical disturbance regimes (ungrazed and burnt) facilitated the invasion of the annual grass species *Bromus tectorum* and other exotic forbs into sagebrush communities (Davies et al. 2009). In comparison, treatments that combined historical and non-historical disturbances (grazed and burnt) did not facilitate an invasion and resulted in greater cover of perennial vegetation. Livestock grazing is a causal factor in *B. tectorum* invasion but can also, depending on the grazing regime, provide resilience to invasion. The mechanism cited here is increased litter accumulation in the ungrazed condition resulting in greater post-fire mortality of native species providing space and resources for invasion. The intensity of grazing determines the outcome because heavy grazing facilitates invasion by decreasing the cover of native plant species. The authors highlight that best management practices are counterintuitive to the assumption that restoring historical disturbances will provide best management of ecosystems.

The best control strategy for the invasive perennial grass *Eragrostis curvula* was the maintenance of grazing (non-historical disturbance) and the use of fertilisers (non-historical conditions) to increase the palatability of the IAP (Firn et al. 2010). Invasion of *E. curvula* is facilitated by selective grazing because of its low palatability in unproductive environments, but the same disturbance, when used strategically, can reduce the population of IAPs. This treatment combination also gave the best results for restoring native dominance. Interestingly, the control strategy of removing the causal disturbance and applying herbicide reduced IAP abundance but led to an increase in abundance of another invasive species, a perennial forb, *Verbena tenuisecta*. In this case, the best approach for the restoration of a novel ecosystem dominated by an invasive species was a combination of non-historical management techniques.

Testable Framework for Strategically Using Grazing to Reduce IAPs in Novel Pastures

The underlying ecological theory behind common control strategies for IAPs is based on traditional successional theory, which, after decades of research, is known to explain the dynamics of few natural systems. Recent studies that explicitly examine the efficacy of control strategies based on traditional models of succession have found that the resultant plant community either remained dominated by the same or another invasive species. Evidence suggests instead that new theoretical models are needed to underpin invasive species control in undesirable novel ecosystems—alternative states models (Wiess 1999; Cox and Allen 2008; Davies et al. 2009; Firn et al. 2010). When these theories are applied to invasion scenarios where ungulate grazing is a relatively recently applied disturbance, and the ecosystem is degraded, manipulating grazing pressure towards the IAP may be a more effective control measure than grazing exclusion (Figure 1) because a threshold to recovery has been crossed (Westoby et al. 1989; Gaertner et al.2012; Richardson and Gaertner 2013). This can be accomplished either directly by changing the timing, frequency and intensity of grazing animals or indirectly by increasing resource availability if the dominant IAP is a plastic species capable of increasing in palatability (Firn et al. 2012).

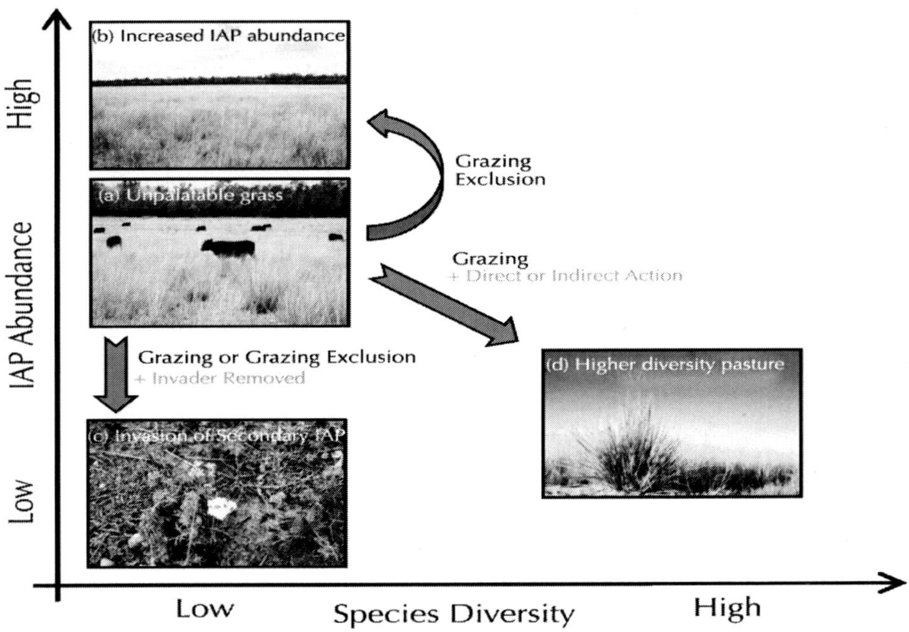

Figure 1: Summary of the results from Firn et al. (2010) on how grazing exclusion and grazing manipulation can alter the abundance of a dominant invasive plant species and small-scale richness of the subsequent plant community. a) Dominant invasive alien plant (IAP) has low palatability and a high growth rate, and the plant community shows low species diversity. b) If grazing is excluded from the site (and not replaced by another disturbance regime), the abundance of the IAP will increase in the short to mid-term and species diversity remains low. c) If grazing is excluded or grazing continued and the dominant invasive plant species killed and/or removed, there is a high probability that a secondary IAP will become dominant. If successful the abundance of the IAP may be low, but species diversity also remains low. d) If grazing is continued but a change is made to the grazing regime, either through direct manipulation of the intensity, frequency and timing of graz-

ing or through indirect manipulation where additional strategies are used to manipulate the palatability of the dominant IAP such as nutrient addition or prescribed burns, then the plant community can be gradually shifted towards a more diverse native species composition.

We know from general ecological theory on the effects of grazing and from invasive species literature that certain preconditions should be met for strategically applied grazing to be a benefit for management (Wiess 1999; Lunt et al. 2007; Cox and Allen 2008; Davies et al. 2009; Firn et al.2010). Native diversity or other more desirable species should be present at the site or in the surrounds as a source of new recruits, and soil moisture and nutrient levels should be adequate for recruiting species to germinate and survive. The decision to use direct or indirect manipulation of grazing will depend on the palatability of the IAP and the economic and technical feasibility of each strategy. Direct manipulation requires decreasing, increasing or maintaining herd numbers, and in some cases intensively guiding their movements. Indirect manipulation will require fertiliser addition, irrigation or prescribed burning. Whether or not these actions are feasible at a site will depend on the characteristics and pressures that are specific to an individual site, landscape and region.

An example of the complexities of using grazing to manipulate an IAP is the establishment of the perennial forb lippia [*Phyla canescens* (Kunth) Greene, native to South America], in the wetlands of Eastern Australia. The relative abundance of this species is affected by grazing as well as by the frequency, season and depth of flooding (Price et al. 2010, 2011a). Further complications are that flooding affects the distribution of grazing animals and therefore their direct and indirect impact on the competing native vegetation of the area (Price et al. 2011b; Whalley et al. 2011). However, there are some indications that high intensity rotational grazing can reduce the abundance of this IAP (Southwell and Reid 2012). Evidence has shown that continuous grazing at dry sites increases bare ground and subsequently increases lippia spread, whereas high intensity rotational grazing at dry and wet sites maintains higher levels of groundcover in general and lower lippia abundance (Figure 2,reproduced with permission from Southwell and Reid 2012). Research by Price et al. (2011a), however, found rest from grazing at different times of the year was not an effective control strategy, probably because stocking intensity was low and did not reduce selectivity, and native species also did not respond to calendar-determined rest periods. Hence, a grazing strategy which reduces selective grazing (high intensity) and its duration, allowing native species a chance at recovery, provided best control of lippia.

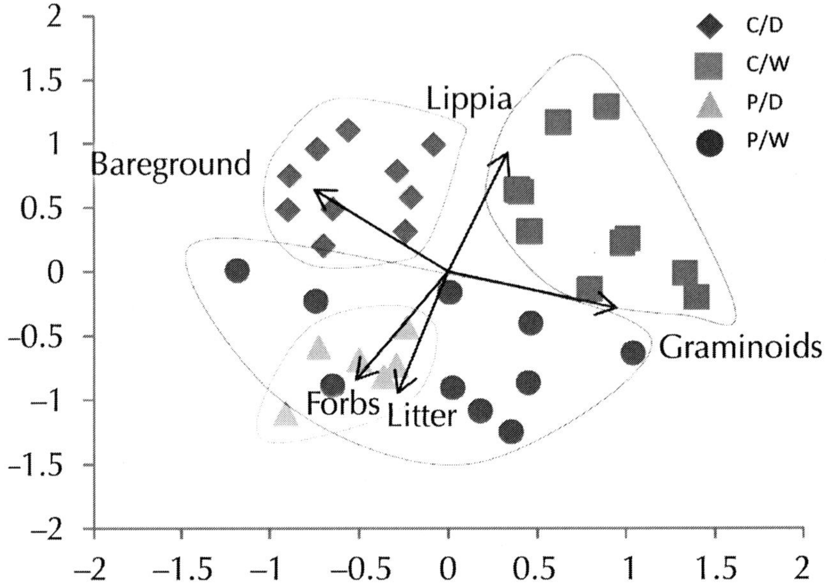

Figure 2: Non-metric dimensional scaling plot for May 2008 sampling period showing changes in invasive alien plant (IAP) abundance for *Phyla canescens* (lippia) under treatments of continuous grazing in dry areas (C/D), continuous grazing in wet areas (C/W), periodic grazing in dry areas (P/D) and periodic grazing in wet areas (P/W). Arrows represent principal component correlation factors > 0.5. Reproduced with permission from Southwell and Reid (2012).

Direct manipulation of grazing will be most effective where the IAP is palatable at certain times of the year, the IAP is not toxic or poisonous to stock, and the IAP demonstrates a different phenology to more desirable species, native or exotic. Other important considerations are whether infrastructure is available for managing targeting grazing such as moveable fencing that will allow for 'cell grazing' or for small paddocks to be created to concentrate stock at the desired locations. In this strategy, grazing herds are forced to remain in certain locations for a given period of time ensuring that IAP abundance is reduced. Although identifying the most appropriate grazing regime is not as simple as targeting the growth and survival of undesirable species, the grazing regime must also reduce negative impacts on the more desirable species. Even if detailed information on phenology is known, finding the best grazing regime is challenging. For example, leafy spurge (*Euphorbia esula*) is an invasive forb that is generally avoided by cattle but eaten by sheep (Rinella and Hileman 2009). Some direct manipulation of grazing pressure has been successful in reducing the density of leafy spurge in rangelands in the USA (Lym et al. 1997) but not in others (Olson and Wallander 1998; Seefeldt et al. 2007). Rinella and Hileman (2009) clarified some of the inconsistencies in leafy spurge research by simulating grazing at different intensities, timing and frequencies. They found that treatments targeting early vegetative stages reduced the IAP and increased resident species, whereas defoliating more intensively at later growth stages often had the opposite effect. These results clearly show that finding the proper timing of grazing is crucial to the success of control efforts.

Indirect manipulation of grazing will be most effective when the palatability of the IAP is highly dependent on resource availability. For example, a highly plastic species, such as *Eragrostis curvula* (Firn et al. 2012), can increase in palatability when a small quantity of fertiliser is applied in a good rainfall year. The key to successfully applying indirect measures for manipulating grazing pressure is that the conditions that make the dominant IAP palatable must also be the same conditions that are most conducive for the recruitment of more desirable species (Firn et al. 2010).

Framework for Managing IAPs with Direct or Indirect Grazing Management

In order to consider the unknowns associated with site-specific variables, we suggest three possible restoration goals to maximise short- and long-term benefits (see Figure 3 for a flowchart summary), and these goals will change considerably based on whether the main land use is farming or conservation. For farming, desired species may include other alien plants that are palatable as the main ecosystem service may be forage value, while for conservation native plants are most likely more desirable. A fundamental issue when using grazing for invasive species management when the goal is conservation and/or to maintain native diversity is when to stop grazing and allow natural recovery to take place. Continuing grazing, whether with a direct or indirect strategy, can result in a community of plant species that are fast-growing and tends to favour the establishment of annuals over perennials, short over tall, unpalatable over palatable, and stoloniferous and rosette over tussock architecture (Diaz et al. 2007). Depending on the disturbance regime of the original community and micro-evolutionary changes that may have occurred in important components, this may not represent the key traits of the previous plant community and may not lead to the control of the IAP. Restoration is a goal-oriented science. For this reason, there must be a set of clearly defined goals for the short and long term. The length of time that a site should be managed under each restoration goal (and the starting point) will depend on the extent to which biotic and abiotic factors of the ecosystem are degraded (e.g. nutrient cycling, hydrology, energy flows, and native species richness). For these reasons, restoration goals should also be flexible so that changes can be made to management strategies when systems do not react according to predictions.

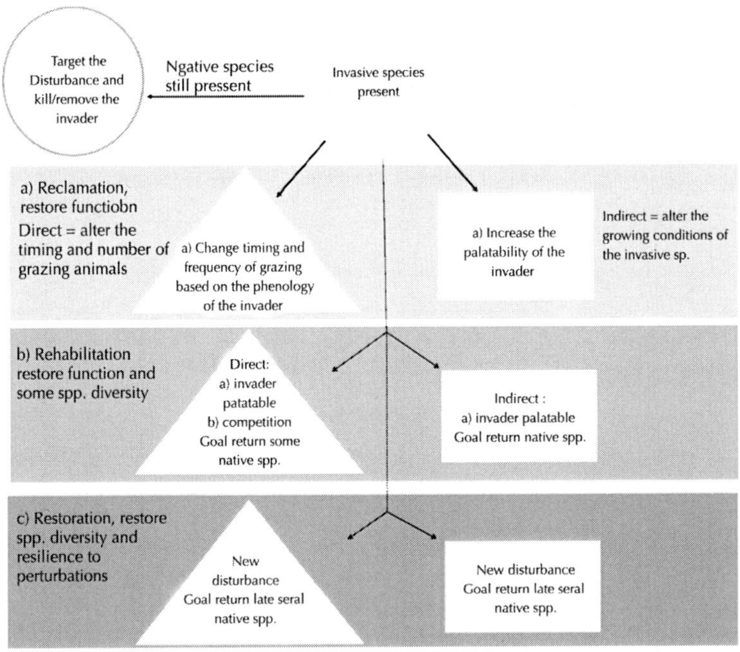

Figure 3: Alternative or a continuum of restoration goals that may help to gradually shift a degraded site, dominated by an invasive plant species, towards a more desirable state with increased species diversity and improved ecosystem functioning. If a grazed pasture has not lost its original ecosystem

functioning, then excluding grazing and removing/killing the IAP may be the best option. If, however, the system has changed to a novel state then direct or indirect manipulation of grazing pressure is a better option. If a grassland community is highly degraded, the main goal may be reclamation. a) At the reclamation stage, the main aim will be to return some form of functioning back to the degraded site. This may be accomplished using either direct or indirect strategies depending on the plasticity of the IAP, economic and social constraints and the phenology of the IAP in comparison to the natives. Evidence that this early stage is working is the return or increase in the abundance of some early seral native species to the site. b) Rehabilitation is where management strategies should aim to maintain key ecosystem functions, while also focusing on encouraging the return of more native diversity to the site. When using direct manipulation of grazing, if the IAP has a different phenology than some of the native species present, then grazing can be applied at a time when the IAP is most susceptible, thereby manipulating competition among species within the community. c) Restoration goals are the hardest to achieve from a management perspective because restoration involves shifting the species diversity from a set of early colonising native species to a system characterised by late seral species. This may not be possible with continued grazing at the site. Instead, what might be needed is a new grazing regime of lower intensity or the application of an alternative disturbance regime, possibly one that resembles more the historical disturbance regime.

When the original species composition is unknown then a potentially useful indicator to assess the 'functional quality' of plant species assembling in response to management in novel ecosystems may be 'soft traits' (defined as easily measured) such as leaf dry matter content (LDMC, mg/g), growth form (e.g. short basal, long basal, semi-basal tussock, erect leafy) and dispersal mode (e.g. anemochory, ballistichory, hydrochory) (Diaz et al. 2004; Garnier et al. 2004; Wright et al.2004; Westoby and Wright 2006). As previously discussed, Firn et al. (2010) found fertiliser increased the palatability of an IAP grass, *Eragrostis curvula*, and a treatment of continuous grazing saw a reduction in its abundance and an increase in more desirable native plant species. This result did not, however, provide any indication as to how the new assemblages of species function and whether they are functioning any differently to the IAP dominated system. Figure 4shows trait diversity (i.e. LDMC, growth form and dispersal modes) versus species diversity (both calculated with Shannon's diversity index) in each of the grazing and fertiliser treatments (unpublished data from the Firn et al. 2010 study). Here, species diversity increased under grazing treatments because the IAP was reduced in abundance. While LDMC decreased, both growth form and dispersal mode diversity increased with the grazing treatments. A decrease in functional diversity of LDMC with grazing suggests plant species are increasing in similarity with increased prevalence of fast-growing resource acquisition specialists, but this treatment opens up opportunities for a more diverse group of species in terms of growth form and dispersal mode to colonise.

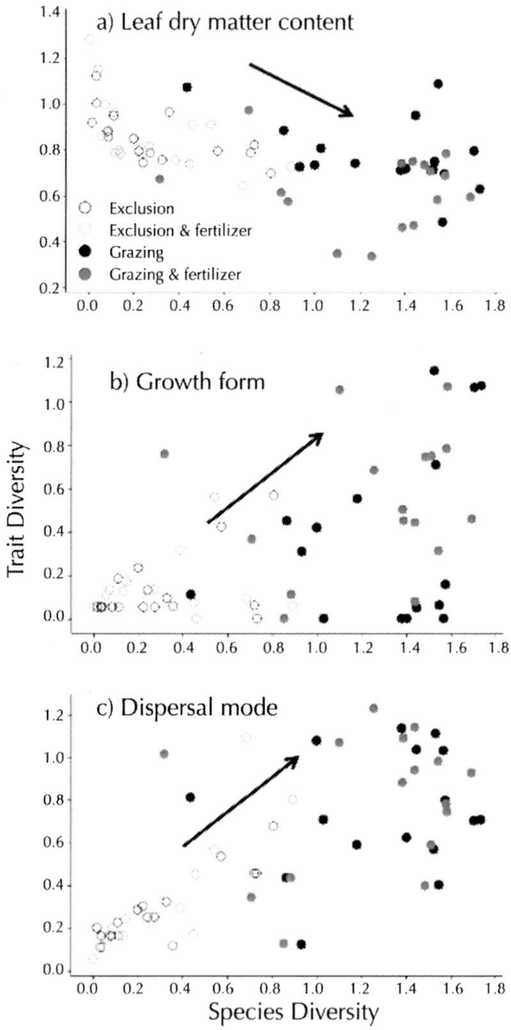

Figure 4. Trait diversity (i.e. leaf dry matter content, growth form and dispersal mode) versus species diversity (both calculated with Shannon's diversity index) across three years of grazing exclusion and fertiliser treatments estab-

lished to reduce the abundance of an IAP, *Eragrostis curvula*(unpublished data from Firn et al. 2010). Panel a) shows Leaf dry matter content,b) growth form and c) dispersal mode. For detailed data collection methods and for species diversity and leaf dry matter content, please see Additional file 1.

Excluding grazing and killing or removing the invasive plant species may be a viable control option at sites where the original functioning of an ecosystem is intact. However, if an ecosystem is found in a new and undesirable state from its historical state, we recommend that the management goal be reclamation. Reclamation is defined as returning some function to a degraded ecosystem generally to ensure continued production (Lamb and Gilmour 2003). At this stage, the prime objective is not likely the return of species diversity, as the system is degraded to such a stage where simply removing the IAP could lead to its re-establishment or the recruitment of another invasive species. Although research has shown that increasing plant species diversity may increase ecosystem services such as productivity, nutrient cycling and resilience to perturbations (Isbell et al. 2011; MacDougall et al. 2013), this is not a realistic goal in the short term. With some invasive species, more intensive management that ensures the continued production of young tissue may be the best option. Evidence that some original functioning has returned could be the recruitment of more desirable species.

Another goal may be rehabilitation, which involves maintaining key ecological functions and encouraging the return of some more desirable species (Figure 1) (Lamb et al. 2005). With this goal, grazing may still be used as a control strategy by focusing management actions on maintaining the grazing pressure on the IAP or by resting grazing at an appropriate time (in terms of phenology) to enable other species to outcompete the IAP (Kemp et al. 1996; Fischer et al.2009). At this stage of the restoration process, the original levels of species diversity will not be returned, but gradually more species may establish. Again these will most likely be early colonists or species that are favoured by disturbance, and the invasive species will likely remain present within the community. For example, Californian serpentine grasslands are renowned for high levels of endemic species that are adapted to low N soils. Nitrogen deposition from cars has provided opportunities for Mediterranean grasses to establish in these grasslands, but cattle grazing can be beneficial by reducing the biomass levels of the IAPs, and trampling reduces thatch that suppresses native species (Wiess 1999). In some cases, where the seed sources of more desirable species have been depleted from the site and surrounding area, enrichment planting or direct seeding may be necessary to encourage species diversity.

The longest-term goal is restoration—re-establishing the species diversity, productivity and structure of the original ecosystem. There is considerable argument as to whether complete historical restoration is an achievable goal (Lamb and Gilmour 2003; Suding et al. 2004; Hobbs et al. 2006) because the original suite of species may not be known or present in the seed bank. Because information on the precise dynamics that governed the original ecosystem is likely also not known, setting targets and milestones for assessing progress is challenging (Lamb and Gilmour2003; Suding et al. 2004). For these reasons, the goal of restoration may need to be altered from the re-establishment of the original suite of species, function and structure to the establishment of late seral species that are favoured by a different disturbance regime other than ungulate grazing. In this case, it may be desirable to exclude ungulate grazing to ensure the colonisation of secondary species not favoured by grazing. Once a diverse set of more desirable species has returned to a site, although overall composition may still include a small population of the IAP, attempting to return the original disturbance regime and excluding grazing may then be the best option for encouraging late seral species.

CONCLUSIONS

There is now strong empirical evidence that the explicit implementation of an alternative states approach for invasive species management can be more effective than control strategies based on traditional models of succession, particularly where grazing is a key mechanism facilitating IAP dominance. Where grasslands are degraded by grazing pressure, the best control strategy is not necessarily to exclude grazing, as is the tendency when biodiversity values are the chief concern. Instead the optimal strategy may be to manipulate the timing, intensity and spatial distribution of grazing to gradually transition the state of the grassland towards a more diverse plant community. It is likely that applying an alternative states approach would be beneficial to other invasion scenarios, but the evidence is not yet available. More research is needed that compares the efficacy of returning historical disturbance regimes and the application of novel disturbances for invasive species control where grazing is not the disturbance, e.g. changes to water flow rates to intermittent wetlands.

Based on current evidence, to explicitly apply this framework, it is essential that the dynamics of the new ecosystem be understood. In a situation where a site is not found in a novel state then excluding grazing could be a viable option. Managing a novel ecosystem to a more desirable state will require a more complex set of management strategies than the generic implementation of common control strategies. To add to this complexity, this approach will also require active monitoring so that the management strategies can be changed over time in response to changes within the ecosystem—adaptive management (McCarthy and Possingham 2007). Based on these requirements, an alternative states approach is more difficult to implement than current generic approaches to IAP management, e.g. herbicides, and may require more effort, money and resources in the short term. In the long term, however, an alternative states approach has the potential to lead to more effective management strategies, even if the historical conditions cannot be recreated, and in this way may prove more cost-effective.

AUTHORS' CONTRIBUTIONS

JF, JNP and RDW collaboratively researched and wrote the review paper. JF collected data shown in Figure 4. All authors read and approved the final manuscript.

ACKNOWLEDGEMENTS

Thank you to Southwell and Reid for granting us permission to use their figure. Jodi Price was funded by the ARC Centre for Excellence for Environmental Decisions.

REFERENCES

1. Ash AJ, Corfield JP, McIvor JG, Ksiksi TS (2011) Grazing management in tropical savannas: utilization and rest strategies to manipulate rangeland condition. Rangeland Ecol Manage 64:223-239.

2. Bendall GM (1973) The control of slender thistle, Carduus pycnocephalus L. and C. tenuiflorus Curt. (Compositae), in pasture by grazing management. Aust J Agric Res 24:831-837.
3. Booth BD, Murphy SD, Swanton CJ (2003) Ecology of weeds. In: Booth BD, Murphy SD, Swanton CJ (eds) Weed ecology in natural and agricultural systems, CABI, Oxford, UK.
4. Cingolani AM, Noy-Meir I, Diaz S (2005) Grazing effects on rangeland diversity: a synthesis of contemporary models. Ecol Appl 15:757-773
5. Clements FE (1916) Plant succession. Carnegie Institute, Washington DC.
6. Cox RD, Allen EB (2008) Stability of exotic annual grasses following restoration efforts in southern California coastal sage scrub. J App Ecol 45:495-504
7. Cummings DC, Fuhlendorf SD, Engle DM (2007) Is altering grazing selectivity of invasive forage species with patch burning more effective than herbicide treatments? Rangeland Ecol Manage 60:253-260
8. Davies KW, Svejcar TJ, Bates JD (2009) Interaction of historical and nonhistorical disturbances maintains native plant communities. Ecol Appl 19:1536-1545
9. De Bruijn SL, Bork EW (2006) Biological control of Canada thistle in temperate pastures using high density rotational cattle grazing. Biol Control 36:305-315
10. Diaz S, Hodgson JG, Thompson K, Cabido M, Cornelissen JHC, Jalili A, Montserrat-Marti G, Grime JP, Zarrinkamar F, Asri Y, Band SR, Basconcelo S, Castro-Diez P, Funes G, Hamzehee B, Khoshnevi M, Perez-Harguindeguy N, Perez-Rontome MC, Shirvany FA, Vendramini F, Yazdani S, Abbas-Azimi R, Bogaard A, Boustani S, Charles M, Dehghan M, de Torres-Espuny L, Falczuk V, Guerrero-Campo J, Hynd A, et al. (2004) The plant traits that drive ecosystems: evidence from three continents. J Veg Sci 15:295-304
11. Diaz S, Lavorel S, McIntyre S, Falczuk V, Casanoves F, Milchunas DG, Skarpe C, Rusch G, Sternberg M, Noy-Meir I, Landsberg J, Zhang W, Clark H, Campbell BD (2007) Plant trait responses to grazing—a global synthesis. Glob Change Biol 13:313-341

12. DiTomaso JM, Brooks ML, Allen EB, Minnich R, Rice PM, Kyser GB (2006) Control of invasive weeds with prescribed burning. Weed Technol 20:535-548

13. Firn J, Rout T, Possingham HP, Buckley YM (2008) Managing beyond the invader: manipulating disturbance of natives simplifies control efforts. J App Ecol 45:1143-1151

14. Firn J, House APN, Buckley YM (2010) Alternative states models provide an effective framework for invasive species control and restoration of native communities. J App Ecol 47:96-105

15. Firn J, Prober SM, Buckley YM (2012) Plastic traits of an exotic grass contribute to its abundance but are not always favourable. PLoS One 7:e35870

16. Fischer J, Stott J, Zerger A, Warren G, Sherren K, Forrester RI (2009) Reversing a tree regeneration crisis in an endangered ecoregion. PNAS 106:10386-10391

17. Gaertner M, Holmes PM, Richardson DM (2012) Biological invasions, resilience and restoration. In: van Andel J, Aronson J (eds) Restoration Ecology - The New Frontier, Wiley-Blackwell, Oxford. pp 265-280

18. Gardener MR, Sindel BM, Whalley RDB, Earl JM (2005) Can *Nassella neesiana*, Chilean needle grass, be incorporated into a grazing management system in Australia? Plant Protect Q 20:36-40

19. Gardener MR, Atkinson R, Renteria JL (2010) Eradications and people: lessons from the plant eradication program in Galapagos. Rest Ecol 18:20-29

20. Garnier E, Cortez J, Billes G, Navas ML, Roumet C, Debussche M, Laurent G, Blanchard A, Aubry D, Bellmann A, Neill C, Toussaint JP (2004) Plant functional markers capture ecosystem properties during secondary succession. Ecology 85:2630-2637

21. Grace BS, Whalley RDB, Sheppard AW, Sindel BM (2002) Managing saffron thistle in pastures with strategic grazing. Rangeland J 24:313-325

22. Gunderson LH (2000) Ecological resilienc— in theory and application. Annu Rev Ecol Syst 4:1-23

23. Hartley MJ, Lyttle LA, Popay I (1984) Control of Californian thistle by grazing management. In: Proceedings of the 37th New

Zealand Weed and Pest Control Conference. New Zealand Weed and Pest Control Society, Hastings, New Zealand. pp 24-27

24. Hayes GF, Holl KD (2011) Manipulating disturbance regimes and seeding to restore mesic Mediterranean grasslands. App Veg Sci 14:304-315

25. Hillebrand H, Gruner DS, Borer ET, Bracken MES, Cleland EE, Elser JJ, Harpole WS, Ngai JT, Seabloom EW, Shurin JB, Smith JE (2007) Consumer versus resource control of producer diversity depends on ecosystem type and producer community structure. Proc Natl Acad Sci USA 104:10904-10909

26. Hobbs RJ (2007) Setting effective and realistic restoration goals: key directions for future research. Rest Ecol 15:354-357

27. Hobbs RJ, Richardson DM (2011) Invasion ecology and restoration ecology: parallel evolution in two fields of endeavour. In: Richardson DM (ed) Fifty years of invasion ecology: the legacy of Charles Elton, Wiley-Blackwell, Malden. pp 61-69

28. Hobbs RJ, Suding KN (2009) Synthesis: are new models for ecosystem dynamics scientifically robust and helpful in guiding restoration projects? In: Hobbs RJ, Suding KN (eds) New models for ecosystem dynamics and restoration, Island Press, Washington DC.

29. Hobbs RJ, Arico S, Aronson J, Baron JS, Bridgewater P, Cramer VA, Epstein PR, Ewel JJ, Klink CA, Lugo AE, Norton D, Ojima D, Richardson DM, Sanderson EW, Valladares F, Vila M, Zamora R, Zobel M (2006) Novel ecosystems: theoretical and management aspects of the new ecological world order. Glob Ecol Biogeograp 15:1-7

30. Hobbs RJ, Higgs E, Harris JA (2009) Novel ecosystems: implications for conservation and restoration. Trends Ecol Evol 24:599-605

31. Holst PJ, Allan CJ, Campbell MH, Gilmour AR (2004) Grazing of pasture weeds by goats and sheep. 1. Nodding thistle (Carduus nutans). Aust J Exp Agric 44:547-551

32. Huwer RK, Briese DT, Dowling PM, Kemp DR, Lonsdale WM, Milchalk DL, Neave MJ, Sheppard AW, Woodburns TL (2005) Can an integrated management approach provide a basis for long-term prevention of weed dominance in Australian pasture systems? Weed Res 45:175-192

33. Isbell FI, Wilsey BJ (2011) Rapid biodiversity declines in both ungrazed and intensely grazed exotic grasslands. Plant Ecol 212:1663-1674

34. Isbell F, Calcagno V, Hector A, Connolly J, Harpole WS, Reich PB, Scherer-Lorenzen M, Schmid B, Tilman D, van Ruijven J, Weigelt A, Wilsey BJ, Zavaleta ES, Loreau M (2011) High plant diversity is needed to maintain ecosystem services. Nature 477:199-202

35. Kemp DR, Dowling PM, Michalk DL (1996) Managing the composition of native and naturalised pastures with grazing. N Zealand J Agri Res 39:569-578

36. Kirby DR, Hanson TP, Krabbenhoft KD, Kirby MM (1997) Effects of simulated defoliation on leafy spruge (*Euphorbia esula*)-infested rangeland. Weed Technol 11:586-590

37. Lamb D, Gilmour D (2003) Rehabilitation and restoration of degraded forests. IUCN and WWF, Gland Switzerland and Cambridge, UK.

38. Lamb D, Erskine P, Parotta J (2005) Restoration of degraded tropical forest landscapes. Science 310:1628-1632

39. Levine JM, Vila M, D'Antonio CM, Dukes JS, Grigulis K, Lavorel S (2003) Mechanisms underlying the impacts of exotic plant invasions. Proc R Soc B Biol Sci 270:775-781

40. Lodge GM, Whalley RDB (1985) The manipulation of species composition of natural pastures by grazing management on the Northern Slopes of N.S.W. Aust Rangel J 7:6-16

41. Lunt ID, Eldridge DJ, Morgan JW, Witt GB (2007) A framework to predict the effects of livestock grazing and grazing exclusion on conservation values in natural ecosystems in Australia. Aust J Bot 55:401-415

42. Lym RG, Sedivec KK, Kirby DR (1997) Leafy spurge control with angora goats and herbicides. J Range Manage 50:123-128

43. MacDougall AS, Turkington R (2005) Are invasive species the drivers or passengers of change in degraded ecosystems? Ecology 86:42-55

44. MacDougall AS, McCann KS, Gellner G, Turkington R (2013) Diversity loss with persistent human disturbance increases vulnerability to ecosystem collapse. Nature 494:86-89

45. Magcale-Macandog DB, Whalley RDB (2000) Genotypic differentiation in *Microlaena stipoides*populations: morphological and ecological patterns. Phil Agri Sci 83:159-172
46. Magcale-Macandog DB, Whalley RDB (2007) Competitive abilities and divergence of *Microlaena stipoides* populations with associated perennial grass species in New South Wales, Australia. Asia Life Sci 16:35-52
47. McCarthy MA, Possingham HP (2007) Active adaptive management for conservation. Conserv Biol 21:956-963
48. Milchunas DG, Sala OE, Lauenroth WK (1988) A generalized-model of the effects of grazing by large herbivores on grassland community structure. Am Nat 132:87-106
49. Noble JC, Whalley RDB (1978) The biology and autoecology of *Nitraria* L. in Australia I. Dist Morphol Potent Utilisation Aust Ecol 3:141-163
50. Olff H, Ritchie ME (1998) Effects of herbivores on grassland plant diversity. Trends Ecol Evol 13:261-265
51. Olson BE, Wallander RT (1998) Effect of sheep grazing on a leafy spurge-infested Idaho fescue community. J Range Manage 51:247-252
52. Olson BE, Wallander RT, Fay PK (1997) Intensive cattle grazing of oxeye daisy (*Chrysanthemum leucanthemum*). Weed Technol 11:176-181
53. Parker JD, Burkepile D, Hay ME (2006) Opposing effects of native and exotic herbivores on plant invasions. Science 311:1459-1461
54. Perring MP, Standish RL, Hobbs RJ (2013) Incorporating novelty and novel ecosystems into restoration planning and practice in the 21st century. Ecol Proc 2013:2-18
55. Piggin CM (1979) Control of *Echium plantagineum* L. with 2,4-D and grazing management. Weed Res 19:17-23
56. Popay I, Field R (1996) Grazing animals as weed control agents. Weed Technol 10:217-231
57. Price JN, Gross CL, Whalley RDB (2010) Prolonged summer flooding switched dominance from the invasive weed lippia (*Phyla canescens*) to native species in one small, ephemeral wetland. Ecol Manag Restor 11:61-63

58. Price JN, Berney PJ, Ryder D, Whalley RDB, Gross CL (2011) Disturbance governs dominance of an invasive forb in a temporary wetland. Oecologia 167:759-769
59. Price JN, Whalley RDB, vanKlinken RD, Duggin JA, Gross CL (2011) Periodic rest from grazing provided no control of an invasive perennial forb. Rangeland J 33:287-298
60. Pywell RF, Hayes MJ, Tallowin JB, Walker KJ, Meek WR, Carvell C, Warman LA, Bullock JM (2010) Minimizing environmental impacts of grassland weed management: can *Cirsium arvense* be controlled without herbicides? Grass Forage Sci 65:159-174
61. Richardson DM, Gaertner M (2013) Plant invasions as builders and shapers of novel ecosystems. In: Hobbs RJ, Higgs EC, Hall CM (eds) Novel ecosystems: intervening in the new ecological world worder, Wiley-Blackwell, Oxford. pp 102-114
62. Richardson DM, Pysek P, Rejmanek M, Barbour MG, Panetta FD, West CJ (2000) Naturalization and invasion of alien plants: concepts and definitions. Diver Distrib 6:93-107
63. Rinella MJ, Hileman BJ (2009) Efficacy of prescribed grazing depends on timing intensity and frequency. J App Ecol 46:796-803 Publisher Full Text
64. Sax DF, Brown JH (2000) The paradox of invasion. Glob Ecol Biogeograph 9:363-371.
65. Scott AW, Whalley RDB (1984) The influence of intensive sheep grazing on genotypic differentiation in *Danthonia linkii, D. richardsonii* and *D. racemosa* on the New England Tablelands. Austral Ecol 9:412-429
66. Seastedt TR, Hobbs RJ, Suding KN (2008) Management of novel ecosystems: are novel approaches required? Front Ecol Environ 6:547-553
67. Seefeldt SS, Taylor JB, Van Vleet S (2007) Reducing *Euphorbia esula* with a combination of sheep grazing and imazapic. J Arid Environ 69:432-440
68. Shea K, Sheppard A, Woodburn T (2006) Seasonal life-history models for the integrated management of the invasive weed nodding thistle *Carduus nutans* in Australia. J App Ecol 43:517-526

69. Southwell M, Reid M (2012) Lippia fact sheet. Central West Catchment Management Authority. Wellington, NSW.

70. Suding KN, Gross KL, Houseman GR (2004) Alternative states and positive feedbacks in restoration ecology. Trends Ecol Evol 19:46-53

71. Thomsen CD, Williams WA, Vayssieres MP, Bell FL, George MR (1993) Controlled grazing on annual grassland decreases yellow starthistle. California Agri 47:36-40

72. Vila M, Espinar JL, Hejda M, Hulme PE, Jarosik V, Maron JL, Pergl J, Schaffner U, Sun Y, Pysek P (2011) Ecological impacts of invasive alien plants: a meta-analysis of their effects on species communities and ecosystems. Ecol Lett 14:702-708

73. Wallace JM, Wilson LM, Launchbaugh KL (2008) The effect of targeted grazing and biological control on Yellow Starthistle (Centaurea solstitialis) in canyon grasslands of Idaho. Rangeland Ecol Manage 61:314-320

74. Waters CM, Murray BR, Melville GJ, Coates DJ, Young A, Virgona J (2010) Polyploidy and possible implications for the evolutionary history of some Australian Danthonieae. Aust J Bot 58:23-34.

75. Waters CM, Melville GJ, Coates DJ, Virgona J, Young A, Hacker RB (2011) Variation in morphological traits among and within populations of *Austrodanthonia caespitosa* (Gaudich.) H.P.Linder and four related species. Aust J Bot 59:324-335

76. Westoby M, Wright IJ (2006) Land-plant ecology on the basis of functional traits. Trends Ecol Evol 21:261-268

77. Westoby M, Walker B, Noy-Meir I (1989) Opportunistic management for rangelands not at equilibrium. J Range Manage 42:266-274

78. Whalley RDB, Price JN, Macdonald MJ, Berney PJ (2011) Drivers of change in the social-ecological systems of the Gwydir wetlands and Macquarie marshes in northern New South Wales, Australia. Rangeland J 33:109-119

79. Wiess SB (1999) Cars, cows, checkerspot butterflies: nitrogen deposition and management of nutrient poor grasslands for a threatened species. Conserv Biol 13:1476-1486

80. Wright IJ, Reich PB, Westoby M, Ackerly DD, Baruch Z, Bongers F, Cavender-Bares J, Chapin T, Cornelissen JHC, Diemer M, Flexas

J, Garnier E, Groom PK, Gulias J, Hikosaka K, Lamont BB, Lee T, Lee W, Lusk C, Midgley JJ, Navas ML, Niinemets U, Oleksyn J, Osada N, Poorter H, Poot P, Prior L, Pyankov VI, Roumet C, Thomas SC, et al. (2004) The worldwide leaf economics spectrum. Nature 428:821-827

81. Yu P, Prakash N, Whalley RDB (2000) Comparative reproductive biology of the vulnerable and common grasses in Bothriochloa and Dichanthium. In: Jacobs SWL, Everett J (eds) Grasses: systematics and evolution, CSIRO, Melbourne.

82. Yu P, Prakash N, Whalley RDB (2003) Sexual and apomictic seed development in the vulnerable grass *Bothriochloa biloba* S.T.Blake. Australian. Aust J Bot 51:75-84

83. Zavaleta ES, Hobbs RJ, Mooney HA (2001) Viewing invasive species removal in a whole-ecosystem context. Trends Ecol Evol 16:454-459

Chapter 5

Scrutinizing the Current Practice of the Environmental Risk Assessment of GM Maize Applications for Cultivation in the EU

Marion Dolezel[1], Marianne Miklau[1], Angelika Hilbeck[2], Mathias Otto[3], Michael Eckerstorfer[1], Andreas Heissenberger[1], Beatrix Tappeser[3], and Helmut Gaugitsch[1]

[1]Umweltbundesamt GmbH, Spittelauer Laende 5, 1090 Vienna, Austria

[2]Ecostrat GmbH, Hottingerstrasse 32, 8032 Zurich, Switzerland

[3]Bundesamt für Naturschutz, Konstantinstrasse 110, 53179 Bonn, Germany

ABSTRACT

Purpose

The prevailing controversies on the potential environmental risks of genetically modified organisms [GMOs] still fuel ongoing discussions among European Union [EU] member states, risk assessors, applicants and scientists, even several years after the commercial introduction of GMOs. The disagreements mainly derive from the current risk assessment practice of GMOs and differences in the perceived environmental risks. Against this background, the aim of this study was to scrutinize the current practice of environmental risk assessment [ERA] of several GMO applications currently pending for authorisation in the EU.

Methods

We analysed the data presented for three assessment categories of the ERA of genetically modified [GM] maize applications for cultivation in the European Union: the agronomic evaluations and the assessments of the effects of GM maize on target organisms and of its potential adverse effects on non-target organisms.

Results

Major shortcomings causing considerable uncertainties related to the risk assessment were identified in all three categories. In addition, two principles of Directive 2001/18/EC are largely not fulfilled - the consideration of the receiving environment and the indirect effects, as mediated, e.g. by the application of the complementary herbicide in the case of herbicide-tolerant GM maize.

Conclusions

We conclude that the current practice of ERA does not comprehensively fulfil the scientific and legal requirements of Directive 2001/18/EC, and we propose improvements and needs for further guidance and development of standards. The recommendations address likewise applicants, risk assessors as well as decision makers.

BACKGROUND

Authorisation of genetically modified organisms [GMOs] in Europe is regulated by Directive 2001/18/EU [1] and Regulation (EC) 1829/2003 [2], respectively. The objective of Directive 2001/18/EC [1] is to protect human health and the environment when a GMO is placed on the market as or in products (Article 1). The focus is on the preventive action due to the ability of living organisms to reproduce and the irreversibility of the release of GMOs in the environment. Thus, the Directive requires a case-by-case approach to the environmental risk assessment [ERA] to be carried out prior to decisions on applications. Annex II of the Directive defines the general principles of the ERA, the steps, the methodology and the conclusions to be drawn on the potential environmental impact from the release or the placing of GMOs on the market. Annexes III and IV lay down the information requirements to be included in applications. When performing the ERA, the applicant has to consider the 'risks to human health and the environment, whether direct or indirect, immediate or delayed, which the deliberate release or the placing on the market of GMOs may pose' (Annex II) [1]. 'Direct effects' are the effects resulting from the GMO itself without an intermediate chain of events. 'Indirect effects' are the effects 'occurring through a causal chain of events, through mechanisms such as interactions with other organisms, transfer of genetic material, or changes in use or management of the crop' (Annex II) [1]. 'Immediate effects' refer to the effects 'which are observed during the period of the release and use of the GMO', whilst 'delayed effects' are the effects 'which may become apparent as a direct or indirect effect either at a later stage or after termination of the release' [1].

Similarly, Regulation (EC) 1829/2003 [2] requires 'a technical dossier for the application of a GMO supplying the information required by Annexes III and IV of Directive 2001/18/EC and information and conclusions about the risk assessment carried out in accordance with the principles set out in Annex II to Directive 2001/18/EC' if the scope of the notification covers food or feed containing or consisting of GMOs (Article 5) [2]. The information requirements as defined by Annex III in the Directive are also referred to in the guidance document of the European Food Safety Authority [EFSA] [3], although they differ to some extent in structure and detail.

For many years, the authorisation of genetically modified plants [GMPs] in the European Union [EU] has been a controversial issue. Regulation (EC) 1829/2003 [2] introduced a centralised procedure of authorisation by the European Commission, which includes the consultation of the European Member States [MS] and a scientific evaluation conducted by EFSA. If a GMO application encompasses cultivation, EFSA nominates a MS to conduct an initial appraisal of the ERA submitted by the applicant. On the basis of this appraisal and complemented with input from MS, EFSA carries out a scientific evaluation and issues an opinion. Based on this opinion, the European Commission drafts a decision to be voted on by MS. Despite the new legislation [1,2,4] and efforts to strengthen risk assessment guidance and to promote exchange among the scientific experts of the EFSA and European MS, no qualified majorities for decisions on import and processing or cultivation of GMPs could be reached so far. With the exception of genetically modified [GM] carnations, all final decisions have been taken by the European Commission unilaterally.

Except for a GM potato in 2010, no GM crops have been approved for cultivation since 1998. Instead, several import and cultivation bans have been enacted by different European MS. For instance, since 1999, the cultivation of *Bacillus thuringiensis* [*Bt*] maize MON810 was banned in Austria, France, Hungary, Luxemburg, Greece and Germany. Since 2009, at the political level, discussions are underway to shift the responsibility for final approval of the cultivation of GM crops to the national level of the individual MS [5].

The difficulties in decision making at the EU level have to be contextualized in light of the controversies among risk assessors, decision makers and scientists in the EU MS regarding the quality and quantity of the data submitted for the legally required ERA of GMPs. On a scientific level, controversies arose on the chosen methodology of the ERA, the tested endpoints, the duration of tests and, last but not least, the chosen test organisms [6-9].

The 'centralisation' of the authorisation procedure and the continuously positive opinions by EFSA considering the ERAs of the applications as sufficient led to a considerable dispute among MS on the presumed safety of GMPs. It has also been criticised by EU MS that the quality of the submitted data varies considerably, and improvements of the ERA based on common standards are being demanded [10-13]. These controversies stress the need for further harmonisation of practices and methods of assessing the environmental risks of GMPs, in particular regarding the impact on non-target organisms [NTOs], potential long-term effects and indirect effects resulting from potential changes in agricultural management techniques.

Against this background, the Environment Agency Austria was commissioned by the German Federal Agency for Nature Conservation to carry out a study with the aim of evaluating the current practice of the ERA of GMPs in the EU, describing the compliance with legal and scientific requirements, identifying possible shortcomings and proposing improvements of the data basis and the ERA process where needed [14]. In this article, the most important findings regarding the evaluation of GM maize applications are presented. More details for GM maize and the results for other GM crops, like herbicide-tolerant [HT] oilseed rape and GM potato with altered starch composition, can be found in the full report [14].

MATERIALS AND METHODS

Selection of Material for Analysis

Seven applications of GM maize including cultivation in their scope, which have been submitted under the relevant EU legislation [1,2] at that time, were selected for this analysis. The selected applications of GMPs comprise traits representative for GMPs of interest for the European market, insect resistance and herbicide tolerance (Table 1). The documents of the original application as well as any additional information submitted by the applicants before 31 December 2007 were accounted for. The selection includes both single-trait events (e.g. maize MON810) and stacked varieties with different combined events (e.g. maize NK603 × MON810).

Table 1: Overview of the applications selected for the study (status: December 2007)

GMP	Relevant legislation	Scope	Application number	GM traits
Maize MON810	D90a, b	Seed (CU, P)	C/F/95/12/02	IR
Maize Bt11	D01b, c	FO, FE, IP, CU	C/F/96/05/10	IR, HT
Maize 1507	D01b	CU, I, Pe	C/ES/01/01	IR, HT
Maize NK603	Rb	FO, FE, I, P, CU	EFSA/GMO/ NL/2005/22	HT
Maize 59122	Rd	FO, FE, I, P, CU	EFSA/GMO/ NL/2005/23	IR, HT
Maize 1507 × NK603	Rd	FO, FE, I, P, CU	EFSA/GMO/ UK/2005/17	IR, HT
Maize NK603 × MON810	Rd	CU	EFSA/GMO/ NL/2005/26	IR, HT

D90, Directive 90/220EEC; D01, Directive 2001/18/EC; R, Regulation (EC) No. 1829/2003; FO, food use; FE, feed use; I, import; P, processing; IP, industrial processing and uses; CU, cultivation; Seed, seed production, HT, herbicide tolerance; IR, insect resistance.

aApplication for renewal for authorisation submitted; bscientific opinion issued by EFSA; csubmitted under Directive 90/220/EEC and transferred to Directive 2001/18/EC; dunder scientific review by EFSA; eexcluding food uses.

Dolezel et al. Environmental Sciences Europe 2011 23:33, doi: 10.1186/2190-4715-23-33

The aim of the study was to evaluate the data basis put forward by applicants for the ERA and its implications for conducting an ERA. This data basis reflects the information available for the Competent Authorities [CAs] of Member States when performing their risk assessment. We point out that the intention of this analysis was not to draw overall conclusions on the potential environmental risks of the GMPs linked with the applications. For this reason and because EFSA issues its opinions after the MS, we decided not to take the EFSA's opinions of the applications reviewed into account. Thus, the analysis exclusively focused on the ERA and the respective data submitted by the applicants in support of their ERA. The data presented for the assessment of effects on human health were omitted from the analysis due to the lack of direct relevance for environmental risks.

Rationale and Analysis of Applications

The procedure of the ERA, including all necessary steps from hazard identification to the overall risk characterisation, is laid down in Annex II of EU Directive 2001/18/EC and the supplementing guidance notes [15]. Moreover, in its Annexes, in particular Annex II.D.2 and Annex III.B, EU Directive 2001/18/EC specifies information requirements which may be necessary to carry out the ERA. Similarly, these information points are required to be included in applications according to Regulation (EU) 1829/2003. In addition, the EFSA guidance document [3] provides specific areas of risk which have to be addressed during the ERA. It is the quality of these information elements which was of main concern in the study commissioned by the German Federal Agency for Nature Conservation [14], whose purpose was to critically analyse the data generation by the applicant in order to perform an ERA and, thereby, the information basis on which decisions have to be made by CAs rather than the methodology of the ERA itself. Consequently, this manuscript does not elaborate on the general methodology of the ERA, its components and underlying concepts, but presents an analysis of the data on which the ERA, the core activity of risk analysis, is based upon.

As the information elements slightly differ in structure and detail between EU Directive 2001/18/EC and the EFSA guidance document [3], we combined them into ten assessment categories: molecular characterisation, expression, agronomics, composition, dissemination and related processes, target organisms, non-target organisms, biogeochemical cycles, change in land use and cultivation technique, and risk management including monitoring [14]. Among the ten categories studied, three selected ones are presented in this article: (1) assessment of the *agronomic characteristics of the GMP*, (2) assessment of *effects on and mediated by target organisms* [TOs] and (3) assessment of *interactions of the GMP with NTOs*.

The analysis of each assessment category followed a common approach. In principle, the analysis focused on studies provided by and carried out on behalf of the applicants. Most of these information are not published peer-reviewed. If the applicant cites or refers to such studies, in a first step, their availability and documentation within the applications was checked. In a second step, these studies were assessed to comply with good scientific practice, including appropriate methodology, experimental design, and statistical analyses and reporting, as well as with criteria laid down in the EFSA guidance document [3]. In accordance with the case-by-case approach outlined in Directive 2001/18/EC (Preamble, points 18 and 19), we considered data on the respective GMP (i.e. the specific transformation event) mandatory for the assessment and thus focused on the data sets specifically generated and presented for the GMP in question. Consequently, only studies conducted with the GMP or with the relevant proteins were further analysed regarding the following aspects:

Agronomic Characteristics of the GMP

Number of field locations; number of years and representativeness of the selected locations with regard to the intended market release (in all cases the entire EU); characterisation of individual locations; choice of comparators (controls); use of appropriate pesticide management in trials with HT GMPs; and relevance of estimated agronomic parameters. The analysis of field data focused on field trials conducted in Europe as these are considered to have the highest relevance if a GMP is to be cultivated in the EU

Effects on and Mediated by TOs

Effectiveness of control; occurrence of secondary pests; effects due to changes in food/prey availability; and resistance development.

Interactions of the GMP with NTOs

Exposure assessment; test strategy and selection of test organism (e.g. assessment of relevant functional groups); toxicological endpoints; multi-trophic effects; and effects on species of conservation concern. In this category, laboratory studies using the introduced protein (e.g. Cry protein), or in some cases parts of the GMP (e.g. pollen), were distinguished from whole-plant studies under containment and whole-plant studies in the field. Additionally, and only for this category, the published literatures cited in the applications as supporting evidence were analysed for their relevance (e.g. type of test substance or GMP used) and their completeness. Under completeness, we checked whether the citations comprised all relevant studies at the time when the application was submitted by the applicant and, thus, whether the state of scientific knowledge was represented.

RESULTS

Analysis of the Assessment of Agronomic Characteristics of the GMP

Data for the assessment of the agronomic performance of the GMP were in all cases generated by conducting field trials at different locations either in one country or in different countries or even continents. Generally, no rationale or criteria were given for the selection of the field sites. For GM maize MON810 and NK603 × MON810, no agronomic data obtained under European conditions were presented (Table 2).

Table 2: Overview of non-European and European field trials for the collection of agronomic data for GMPs

GMP	Years: non-European countries	Years: European countries/no. of locations	Herbicide applied (European trials only)
Maize MON810	-	-	-
Maize Bt11	-	1995: FR/-	glu-
Maize 1507	-	2000: FR+IT+BG/3+2+1	glu+/glu-a
		2002: ES/3	glu-
Maize NK603	1999: US	2000-2001: DE/4	gly-
		2002: FR/4	gly-
Maize 59122	2002/2003: CL	2003: BG/3	glu+/glu-b
	2003: US, CA	2004: BG+ES/3+3	

Maize 1507 × NK603	-	2003: BG+ES/2+3	gly+ fb glu+
Maize NK603 × MON810	2002: US	-	-

Two-letter country codes refer to ISO 3166-1 alpha-2 codes (IT, Italy; ES, Spain; FR, France; BG, Bulgaria; DE, Germany; CA, Canada; CL, Chile; US, United States). '-', no data/information provided in the application; glu+, treatment with glufosinate; glu-, no treatment with glufosinate; gly+, treatment with glyphosate; gly-, no treatment with glyphosate; fb, followed by.

aNo statistics applied; bagronomic parameters were assessed in non-treated plants only.

Dolezel et al. Environmental Sciences Europe 2011 23:33, doi:10.1186/2190-4715-23-33

Information on field trial locations was generally restricted to stating the name of the location and the country. Climatic data were presented in the applications of GM maize 1507, 59122 and 1507 × NK603; for maize 59122 and maize 1507 × NK603, information on the soil type at a specific location and a map of the location were also presented (see [14]).

In a given year, field trials were conducted in one (1995; maize *Bt*11; 2002: maize 1507; 2000-2002: maize NK603; 2003: maize 59122) or two (2004: maize 59122; 2003: maize 1507 × NK603) or three (2000: maize 1507) European countries (Table 2). Except for GM maize 59122 in Bulgaria, none of the field trials was carried out for more than two consecutive growing seasons at a given location.

In 60% of genetically modified herbicide-tolerant (GMHT) maize applications with European field trials, the agronomic performance was tested in the absence of the corresponding broad-spectrum herbicides. For GM maize 1507 and 59122, the respective herbicide treatments were included (Table 2) in the field trial design. However, for GM maize 1507, the potential differences between the herbicide-treated GM maize and the untreated controls were not statistically evaluated in one year (2000), and no herbicide treatment was included in the other year of field trials (2002). For GM maize 59122, agronomic parameters were assessed in non-treated GMPs in all years and at all sites.

The assessment of the agronomic performance of the GMP included plant growth and development, plant morphology, plant health and yield using a broad range of different parameters. Both, quantitative (e.g. ‹yield›) and qualitative (e.g. ‹leaf colour or shape›) parameters, were assessed by the applicants (see [14]). The measured parameters were not consistent across all GM maize applications.

Insect damage of the GMP was evaluated in 29% of GM maize applications with field trials in Europe (Table 3). No evaluation of insect damage under European field conditions was carried out for maize events MON810 and NK603 × MON810. In the case of maize *Bt*11, assessment of insect damage in one European country was stated, but no data were presented in the application.

Table 3: Insect damage evaluated in European field trials for the assessment of agronomic traits of the GMP

MP	European Field trial (year/country)	Parameter assessed	Method	Insect pests assessed
Maize MON810	-	-	-	-
Maize Bt11	1995/FRa	-	-	-
Maize 1507	2000/FR, IT, BG	Infestation	1-9 scale, v. e.	-
	2002/ES	Infestation	1-9 scale, v. e.	-
		Insect tunnels in stalks	Insect tunnelling length (cm)	O. nubilalis, Sesamia sp. (2 of 3 sites)b
Maize NK603	2000/DE	Infestation	1-6 scale, v. e.	O. nubilalis
		Difference to other insects	Yes/no class.	-
	2001/DE	Infestation	1-6 scale, v.e.	O. nubilalis
		Difference to other insects	Yes/no class.	-

	2002/FR	Abundance of larvaeb	Larvae per stalk/ear	O. nubilalis (4 sites) Sesamia sp. (3 sites)
		Borer tunnelsc	Length of borer tunnelling in stalk/ear	
		% damaged earsc		
Maize 59122	2003/BG, 2004/ES, BG	Infestation	1-9 scale, v. e.	-
Maize 1507 × NK603	2003/ES, BG	Infestation	1-9 scale, v. e.	-
Maize NK603 × MON810	-	-	-	-

The infestation rates of pests were evaluated by visual observation and a qualitative rating on a 1 to 6 or a 1 to 9 scale. In the applications of GM maize 1507 and NK603, additionally, quantitative methods were used, such as assessing the length of the insect tunnels in the maize stalks or ears (Table 3).

For insect-resistant [IR] maize applications with field trials in Europe, the pest species assessed in the insect damage recording were specified only in the case of maize 1507, although not in all years. For this IR maize and also maize NK603, although the latter conferring herbicide tolerance only, the infestation rate specifically caused by the European corn borer (*Ostrinia nubilalis*) was evaluated. In these two applications, also the Mediterranean corn borer or pink borer (*Sesamia*sp.) was evaluated, but also not at all sites. In no case was reference made to the pest insect pressure of the respective year and location. With the exception of maize 1507, it was never indicated whether the background infestation levels of the TOs in the control plants were sufficient to demonstrate a relevant level of control by the GMP.

Analysis of the Assessment of Effects on and Mediated by Target Organisms

In applications of HT maize, no TO (i.e. target weeds) was identified. TOs (i.e. target pest insects), however, were addressed in applications of IR maize. In applications of IR maize expressing the Cry1Ab toxin (maize MON810, maize *Bt*11, maize NK603 × MON810), *O. nubilalis* and *Sesamia*spp. were indicated as TOs. The applicant of *Bt*11 maize listed also other lepidopteran maize pests occurring in the UK, such as *Acronicta rumicis*, *Agrotis* spp. and *Autographa gamma*, as TOs. For maize expressing the Cry1F toxin (maize 1507, maize 1507 × NK603), *O. nubilalis* and *Sesamia*spp. were listed as TOs. For maize 59122 expressing a binary toxin (Cry34Ab1/Cry35Ab1), the Western corn rootworm (*Diabrotica virgifera virgifera*) was indicated as the sole TO of this GMP in the EU.

In 67% of IR maize applications, data on the specificity of the respective Cry proteins to TOs were provided (Table 4). However, in these cases, the applicant provided data produced to demonstrate the equivalence of the plant-expressed proteins to bacteria-expressed proteins, for instance by determining bioactivity. Insect bioassays on pest species were used to conclude on the bioactivity. Data from bioassays using the actual TOs and the respective GM maize plant material for which approval was sought were not provided, with the exception of maize 1507. However, we consider the efficacy data for maize 1507 as not scientifically robust due to the fact that the demonstration of the capability of 1507 maize to control TOs rested on a study presenting two mortality values from a single, non-replicated study at one location in one year in addition to one study, which did not fulfil minimal scientific requirements in terms of sample size, replication and statistical analysis. Studies showing reliable and sustainable control of *Sesamia nonagrioides* have not been presented by the applicant of 1507 maize, and robust data on the degree of mortality related to different growth stages and tissues of Cry1F maize were not presented (see [14]).

Table 4: Studies evaluating the efficacy of the GMP in IR GM maize applications

GMP	Cry protein expressed in GMP	Test substance used in ERA	Pest species evaluated in ERA
Maize MON810	Cry1Ab	-	-
Maize Bt11	Cry1Ab	Plant and bacterial proteins	ECB, CEW
Maize 1507	Cry1F	1507 maize plants	ECB, MCB
	Cry1F	Plant and bacterial proteins	ECB, CEW, TBW, FAW, BCW
	Cry1F	1507 maize plants	MCB
Maize 59122	Cry34Ab1/ Cry35Ab1	Bacterial proteins	SCR
	Cry34Ab1/ Cry35Ab1	Bacterial proteins	CEW, ECB, BCW SCR, WCR, NCR, CLA
	Cry34Ab1/ Cry35Ab1	Bacterial proteins	CEW, ECB, NCR, WCR, CBW, CLA
Maize 1507 × NK603	Cry1F	See maize 1507	See maize 1507
Maize NK603 × MON810	Cry1Ab	-	-

'-', no specific data provided by the applicant; ECB, European corn borer (Ostrinia nubilalis); CEW, corn earworm (Helicoverpa zea); MCB, Mediterranean corn borer (Sesamia nonagrioides); TBW, tobacco budworm (Heliothis virescens); BCW, black cutworm (Agrotis ipsilon); FAW, fall armyworm (Spodoptera frugiperda); SCR, Southern corn rootworm (Diabrotica undecimpunctata howardi); WCR, Western corn rootworm (Diabrotica v. virgifera); NCR, Northern corn rootworm (Diabrotica barberi); CLA, corn leaf aphid (Rhopalosiphum maidis); CBW, cotton boll weevil (Anthonomus grandis).

Dolezel et al. Environmental Sciences Europe 2011 23:33, doi:10.1186/2190-4715-23-33

The applications of the stacked events contained no studies with the GMPs in question and cited the studies submitted for the respective single-event IR maize application described above. The potential development of resistance of TOs was taken into account in all applications of IR GMPs. In each case, insect resistance management plans were included for *O. nubilalis* (GM maize expressing the Cry1Ab or Cry1F toxins) and for *Diabrotica* spp. (GM maize 59122). No resistance management plan was provided or discussed for *Sesamia* spp. in the Mediterranean region or for any other listed target insect organism. For HT GMPs, potential adverse environmental effects resulting from the development of resistant weeds were not considered.

Analysis of the Assessment of Interactions of the GMP with Non-Target Organisms

All applicants referred to studies assessing the exposure and/or effects of the respective GMP on NTOs. However, the number of studies and their quality varied significantly among the applications (Table 5). In none of the GM maize applications was a specific evaluation of the different exposure pathways of different NTOs in different environmental media (e.g. vegetation, soil and water) conducted. In 43% of the applications, the exposure of some organisms or groups of organisms to the GMP was discussed in the ERA (Table 5). In the application of maize *Bt*11, published studies describing the exposure of Monarch butterflies to *B. thuringiensis* maize in the USA were cited, and the applicant referred to published and unpublished studies on the effects of the Cry1Ab toxin on Lepidoptera and non-lepidopteran NTOs. No specific exposure values for individual species, regardless whether in Europe or elsewhere, were calculated or assessed in any detail for maize*Bt*11.

Table 5: Studies assessing exposure of NTOs and the adverse effects of GMPs on NTOs

GMP	Exposure assessed in ERA	Lab studies	Greenhouse studies	Field studies	Tritrophic studies	Species of conservation concern/ aesthetical/ cultural value

Maize MON810	-	Y	-	Y	-	-
Maize Bt11	Y	Y	-	Y	-	- (M, O)e
Maize 1507	Y	Ya	-	Y	-	Y (M)
Maize NK603	-	Y	Y	-	Y	-
Maize 59122	Y	Y	-	Y	Yd	Y (M)
Maize 1507 × NK603	-	Y, Nb	-	Y	-	-
Maize NK603 × MON810	-	Y, Nc	-	-	-	- (M)

Only field studies with the respective GMP were considered.

Lab studies, with isolated proteins/parts of the GMP; greenhouse studies and field studies, with whole plants; Y, study conducted by the applicant or by order of the applicant; N, reference to assessment in other applications;'-', no data presented or applicant referred only to the published literature; M, Monarch butterfly; O, other.

aThree of the nine studies cited are missing in the application; breference to a single-event application of maize 1507; creference to single-event applications of maize MON810 and maize NK603; dstudy not attached to the application; ereference to published studies, no toxicological studies submitted by the applicant.

Dolezel et al. Environmental Sciences Europe 2011 23:33 doi:10.1186/2190-4715-23-33

In the application of maize 1507, exposure was evaluated for non-target Lepidoptera by estimating the predicted environmental concentration based on pollen expression levels. The sensitivity data (LD_{50}) were calculated from toxicological studies of 16 lepidopteran species which, with the exception of the Monarch butterfly, were all pest species [16]. For maize 59122, the *B. thuringiensis* toxin concentration in different plant tissues and organisms was calculated and a 'formal' exposure assessment according to the US EPA Guidelines for the ERA was carried out.

The effects of the toxin expressed in the GMPs on NTOs were assessed at different levels of containment (laboratory, greenhouse or field). In 100% of the applications, the effects of *B. thuringiensis* proteins on NTOs were evaluated in laboratory studies using a standard set of test species representing a typical toxicological testing regime, e.g. as used for pesticide testing. Most testing was carried out using honeybees, earthworms, green lacewings, parasitic hymenoptera and ladybird beetles (see [14]). Generally, the isolated bacteria-expressed proteins were fed to the test organisms. Additionally, pollen (e.g. ladybird beetles, honeybees, daphnids); protein suspensions (honeybees, adult ladybirds and parasitic hymenoptera); moth eggs coated with bacteria-derived protein solutions (green lacewings); or soils amended with bacteria-produced proteins (earthworms, springtails) were used. The same laboratory toxicity tests were submitted in different applications of different *B. thuringiensis* maize expressing the same class of toxins (e.g. Cry 1Ab) without taking into account possible event-specific differences. The measured parameters related mostly to mortality and, therefore, acute toxicity. The endpoints used were no-observed-effect-concentration [NOEC], LC_{50} or unspecified 'signs of toxicity'. Depending on the species assessed, sublethal parameters were measured in 0% (parasitic hymenoptera, fish), 20% (bees), 25% (pest Lepidoptera), 33.3% (daphnids, green lacewing larvae), 40% (earthworms, coccinellids), 50% (carabids, aphids), 80% (collembola) and 100% (mammals, birds and flower bugs) of the laboratory studies with NTOs presented in the applications [14].

Only data on NK603, a GMHT maize, were generated in the semi-controlled greenhouse experiments. None of the IR maize was equally evaluated under greenhouse conditions (see [14]). In 57% of the applications, the GMP was assessed in field studies for potential non-target effects under European conditions (Tables 5 and 6). The plots used in these field studies were generally small or contained only a few plants (see [14]). Sampling of NTOs was either done using different types of traps (pitfall or sticky traps, beating of plants) or by visual inspection of the plants. Either the abundance of the organisms was recorded or a simple inventory list of species compiled.

Table 6: Studies assessing the adverse effects of GMPs on NTOs under field conditions

GMP	US: Locations (years)	EU: Locations/country (years)
Maize MON810	2 (1993); 2 (1994); 3 (1995)	2/FR (1995)

Maize Bt11	4 (1997-1998), 2 (1998-19999), 4 (1998-1999), 2 (2000-2001)	-
Maize 1507	1 (1999)	2/ES (2002); 1/FR (2000)
Maize NK603	-	-
Maize 59122	2 (2001, 2002)	1/ES (2006); 1/HU (2006); 2/ES (2005)
Maize 1507 × NK603	-	1/ES (2005)
Maize NK603 × MON810	-	-

Only field studies with the respective GMP were considered.

'-', no study submitted

Dolezel et al. Environmental Sciences Europe 2011 23:33 doi:10.1186/2190-4715-23-33

Tritrophic studies were provided in two (29%) of the applications (GM maize lines 59122 and NK603). However, the study provided for GM maize 59122 lacked a verification of the toxin uptake by the prey and the natural enemy species, respectively. Species of conservation concern or of aesthetical and cultural value were considered in 57% of the applications (Table 5). In all these cases, the Monarch butterfly (*Danaus plexippus*) was considered. For maize *Bt*11, the applicant additionally cited Lepidoptera listed in Annex IV of the Habitats Directive 92/43/EEC without providing any toxicity data or detailed risk assessment.

Potential secondary pests, i.e. pest species replacing the target organism and being largely insensitive to the respective toxin, were not considered in any of the applications. For maize *Bt*11, secondary infestations of maize by pathogens such as *Fusarium* sp. were discussed. The applicants of the two stacked event maize applications referred to the risk assessments submitted in the single-event applications. As for TOs, no further tests with NTOs and the stacked GMP event or the combination of the bacteria-derived toxins expressed in the stack were conducted.

Of the GM maize applications, 57% contained studies carried out by the applicant with the respective GMP under field conditions in Europe

(Table 6). European studies were conducted in one or two countries at one or two locations per country. In two cases, field studies were mentioned, but data of the respective studies were not shown (maize NK603 and NK603 × MON810). Additionally, applicants referred to published field studies on the NTO effects of *B. thuringiensis*crops. However, the respective GMP used in the literature was generally not indicated when the applicant cited published studies for the discussion of the potential adverse effects of the GMP on NTOs. Applicants also referred to studies in which not the respective GMP that was the subject of the requested approval was used. For example, for *Bt*11 maize, the applicant presented a summary report of various surveys of NTOs in *B. thuringiensis* maize fields. This report was not specifically compiled for *Bt*11 but for the registration of *B. thuringiensis* corn by US authorities and consequently contained studies with different *B. thuringiensis* maize events. Five of the 16 studies were conducted with *Bt*11 maize. From the total of 16 studies, five were conducted in Europe, however none of these with maize *Bt*11. Concerning maize NK603, seven field studies were cited, but six of these dealt with other GM crop species such as HT soybean or HT wheat. The studies cited for stacked event maize applications generally dealt with the single-event GMPs.

DISCUSSION

From our analyses of seven GM maize applications, we conclude that the submitted data for ERA are incomplete and many crucial aspects regarding ecology and biostatistics are not addressed. Interactions between the GMP and its receiving environments are not sufficiently taken into consideration, and only a few field trials were carried out in Europe. Furthermore, in the field trials conducted in Europe, few aspects relevant to the ERA were included. The field assessments focussed mainly on agronomic rather than on ecological parameters. A range of consequences for the different components of the ERA result from these shortcomings. The most important ones will be discussed in the following sections.

Agronomic Evaluations of the GMPs Lack Scientific Depth and Standardized Protocols

The methods and parameters used for the evaluation of the agronomic characteristics of the GMPs varied considerably between the applications. We propose to use standardized parameters and/or methods for the assessment of any potential agronomic differences between the GM maize and its non-GM control. This is particularly relevant for the assessment of insect damage. The assessment parameters 'difference to other insects' or 'infestation', as employed in many applications, cannot be considered as sufficient for the demonstration of the efficacy of the GMP regarding different insect pests. The assessed insect pest species were often not specified, and rarely quantitative assessments with the respective GMP were made. In addition, some of the GM maize applications entirely lacked evaluations of pest susceptibility under European conditions.

A standardized methodology to assess potential differences between the GMP and the control plant in the susceptibility to common and locally important pest species is urgently needed for applications of GM maize with IR traits. Existing methods for the evaluation of pest infestation applied in the evaluation of plant varieties or the authorisation of plant protection products may also be helpful for reporting and evaluating relevant pests of GMPs [17-20]. For GMPs, such methods should be specifically adapted or developed to give applicants a clear guidance.

Definition of Target Organisms Not Consistent with Insect Resistance Management Plans

Regarding the assessment of effects on and mediated by TOs in the applications, the criteria applied to define TOs for IR maize remain unclear. For example, in some cases, all maize-feeding Lepidoptera were considered TOs. Defining the 'target' (and 'non-target') pests for a specific insect-resistant GMP is not trivial. A lepidopteran-active *B. thuringiensis* toxin may work well only against some out of several lepidopteran pest species that co-occur on the same crop. For instance, in the USA, the GM maize varieties expressing the Cry1Ab toxin are most effective against the European corn borer, but show lesser insecticidal activity against *Spodoptera* spp. For pest control purposes, *Spodoptera* species are not necessarily 'targets' because they are not reliably controlled below a damage threshold. In our view, TOs should include only those species that are aimed at being controlled by the respective GMP and for which the efficacy of the IR crop has been demonstrated. All others should be classified as non-target pest species.

Therefore, defining the TO for insect-resistant GM crops should clearly distinguish which lepidopteran pests are targeted in combination with the proven efficacy of the GMP for each specific pest. This should include a delineation of pest species which are not the primary targets of the GMP but may or are known also to be affected by the GM trait. The potential of the development of new major pests and resistance development in these species should also be included in the evaluation. This is necessary for reasons of quality control, product liability and resistance management.

The identification of the TO(s) for a particular *B. thuringiensis* maize event by the applicants was not consistent with the submitted insect resistance management [IRM] plans for *B. thuringiensis*crops. These were exclusively focused on *O. nubilalis* or *D. v. virgifera*, respectively. Other pest species, even if listed as TOs, such as *Sesamia* spp., were always left unconsidered in IRM plans. Especially lepidopteran pest species that may be affected to a lesser degree by *B. thuringiensis*toxins (e.g. *Spodoptera* spp., *Agrotis* spp.) [21,22] might develop resistance fairly quickly. A pest species may be classified as a 'non-target pest' for pest control purposes if it is not reliably controlled by a particular *B. thuringiensis* crop (e.g. *Spodoptera* spp.), but as a 'target' from the perspective of the insect resistance manager if it is sublethally affected. Hence, a careful definition and justification of what is a target and a non-target pest is important as it has implications for the evaluation of the interactions of GMPs with NTOs.

Efficacy of the GMP Not Tested in the Field

Efficacy data are a basic element in the assessment of effects on and mediated by TOs and are as well important for the evaluation of the interactions of GMPs with NTOs. For *B. thuringiensis* maize expressing the Cry1Ab or Cry1F protein, only in 20% of the applications were the target pest species *O. nubilalis* and *Sesamia* sp. evaluated in the field (maize 1507; Tables 3 and 4). For maize 59122, no field data were presented providing evidence that this GM maize is effective in the field against its coleopteran TO as the insects evaluated in the field were not indicated in the application (Table 3). Hence, there is no information whether maize 59122 is actually controlling its TO under representative field conditions in Europe.

In our view, especially with IR crops, the abundance of not only the target but also non-target pest species should play an important role in the selection of the field trial locations considering different GM crop performances due to differential pest infestations and damaging levels [23-25]. Reporting local target pest densities is also relevant for understanding the baseline infestation rates in the respective regions. As insect or weed pressure may vary depending on regional characteristics, local conditions or years, these must be considered as well. No information was provided in the applications as to whether the background infestation levels of the TOs in the control plants were sufficient to demonstrate the efficacy of the product. This is important not only for the quality control of the GMP but also for balancing the positive desired effect of the treatment (i.e. the TO control) against any potential adverse environmental effects. If no sufficient control under certain agronomic conditions can be demonstrated, it is questionable whether the risk for resistance development or of the adverse environmental effects of the GMP is acceptable.

For example, plant protection products have to be tested under relevant circumstances where the TO has been shown to be present at a level causing adverse effects on an unprotected crop [26]. We suggest gaining the necessary data in the course of the agronomic evaluation of the GMP in relevant and representative locations of the different receiving environments. This would allow judging whether or not the GMP is effective in the envisaged cultivation area and under different agronomic conditions. Moreover, it would allow the identification of those production regions with the highest pest pressure.

In addition, there is an urgent need for the development of standardized procedures that specify which tissues and growth stages of the GMP for each relevant target pest should be tested. We recommend drafting respective guidance for IR GMPs. Guidelines for plant protection products which could be adapted to GMPs are currently available [19].

Efficacy of the GMP Not Adequately Assessed In the Laboratory

For the assessment of effects on and mediated by TOs, data on the efficacy and on the basic biological effects (e.g. mortality, development time, fecundity, etc.) of the GMPs for all listed target pests are essential. In the laboratory studies submitted in order to demonstrate the efficacy of the*B. thuringiensis* maize events (Table 4), applicants generally tested the bacteria-produced GM proteins instead of the GM plant. By using only purified bacterial proteins in optimized diets, any plant/toxin interactions are excluded, the potential bioactivity of smaller fragments of *B. thuringiensis* proteins, e.g. as found in maize MON810, are overlooked [27], and the variability in toxin concentrations *in planta* [28] is not taken into consideration (see also [7]).

Additionally, these lab studies often included pest species that do not occur as pests in Europe (e.g. corn earworm *Helicoverpa zea*, tobacco budworm *Heliothis virescens*, fall armyworm*Spodoptera frugiperda*). In all but one case of IR maize expressing a lepidopteran-active toxin (maize 1507), the applicants failed to evaluate the efficacy of the respective proteins to *Sesamia*sp., a pest common in the south of Europe. Only 67% of the studies for maize expressing a coleopteran-active toxin (maize 59122) assessed the relevant TO occurring in Europe, the Western corn rootworm, *D. v. virgifera*. For maize MON810 and the corresponding stacked maize NK603 × MON810, neither studies with *O. nubilalis* nor *Sesamia* spp. were submitted.

Non-Consideration of the Required Broad-Spectrum Herbicides and Their Potential Effects on the Receiving Environment

TOs were only identified for GMPs with IR traits, but not for HT traits. As a consequence, the effects of the GMP-herbicide complex on weeds were not assessed in any of the applications. However, the effects on weeds are relevant for the evaluation of the interactions of the GMP with NTOs. In our view, this reflects one of the largest deficiencies of the current regulation of HT GMPs - the exclusion of adverse effects caused by the altered management and use of the corresponding broad-spectrum herbicide. To our understanding, this is in clear contradiction to the requirements of Directive 2001/18/EC recommending an assessment of such indirect effects[1,15].

As a consequence, potential differences in the agronomic or environmental performance of HT maize compared with conventional maize due to the application of the non-selective herbicide were not addressed in any of the applications. Data on the performance and efficacy of the HT GMP under field conditions, for instance an evaluation of whether GMHT maize would require different herbicide regimes at different times and the consequences of that for the flora and fauna in agroecosystems, were not submitted. The Farm Scale Evaluations demonstrated that herbivores, detritivores, and many of their predators and parasitoids in arable systems are sensitive to the changes in weed communities that result from the introduction of new herbicide regimes [29-32]. Changes in herbicide use are known to be major driving forces for the intensification affecting biodiversity in European agricultural landscapes [33-35] (see also [36] for GMHT crops). In this respect, it has to be emphasised that weed communities differ significantly between continents, countries or even agricultural regions; hence, any evaluation should consider regional differences and include plant species growing in the agricultural environment being of national or regional conservation concern. However, none of these were taken into consideration in GMHT maize applications. Also, resistance development in weeds was left unconsidered within the ERAs. This is in contrast to the GM maize with IR traits where the resistance development of TOs has been considered as the most important (agronomic) risk associated with the cultivation of such GMPs.

Applicants generally shift the responsibility for the assessment of potential adverse environmental effects due to the use of the complementary herbicide to the authorisation of the herbicide according to Directive 91/414/EEC [26]. We believe that considerable improvements in the ERA of GMHT crops could be made if applicants would include some information on the relevant plant protection product to be used with the GMHT crop submitted according to Directive 91/414/EEC. This concerns in particular details on the application of the plant protection product (rate, timing, instructions for use, etc.), information on the toxicological profile including potential metabolites and data on the efficacy of the product including resistance risks. This information should also be available to authorities under Directive 2001/18/EC when cultivation of a GMHT crop is envisaged. Otherwise, an assessment of changed management procedures as recommended in Directive 2001/18/EC cannot be conducted.

Inadequate Approach for the Evaluation of Interactions of the GMP with NTOs

The current approach for the assessment of the effects on NTO applied by the applicants in the applications reviewed is based on the assumption that a GMP consists of two parts that function in a linear additive fashion: the crop and the novel GM transgene product. Based on the substantial equivalence concept and when no statistically relevant compositional changes are detected, the crop plant as such is declared as safe and, consequently, only the added transgene product is subject to testing in the ERA. This approach to the assessment of NTOs is highly disputed in the scientific community [6,37-40] and fails to comply with the provisions put forward by Directive 2001/18/EC [41,7]. A consequence of this interpretation of the legal and scientific requirements for ERA is that clear risk research hypotheses are often not formulated and tested for the GMP in question. Hence, the assessment lacks a well-founded exposure analysis of the GMP in its receiving environments and meaningful testing of the selected NTOs from these environments.

In the reviewed applications, only a few applicants carried out an exposure assessment for the respective GMP (Table 5). The expression values used were either not based on values derived from European field trials or were considered in pollen only. For example, the exposure values of maize 59122 were based on the expression values of one toxin only (Cry34Ab1), although the pesticidal action of this event depends on the binary toxins Cry34Ab1/Cry35Ab1 [42]. Once grown commercially, GMPs and GMP transgene products are released into the environment and may affect target and NTOs via different pathways. Even after harvest, the novel proteins will remain in different environmental compartments and affect organisms in different environmental media or at different trophic levels [43-47]. A well-founded assessment of potential exposure routes constitutes an important first step in this ERA category. As for the regulation of plant protection products according to Directive 91/414/EEC, an analysis of the distribution, fate, and behaviour of the GMP and parts of the GMP, together with its transgenic products in the environment, should be self-evident also for GMPs.

The assessment of the potential effects of the GMP on NTOs generally comprised few standard laboratory trials with the isolated bacteria-derived GM protein. Ecotoxicity testing of the protein in the absence of the plant context has been discussed in detail in the full report [14] and by other authors [7,48,49]. Neither greenhouse studies nor tritrophic studies were consistently provided, thus not taking any risks to predators of herbivores feeding on GMPs into consideration. The few field studies provided for evaluating potential adverse effects on NTOs were designed as stand-alone trials with no connecting research hypotheses linking laboratory and/or greenhouse studies with field studies.

Assessment of the potential adverse effects of a GMP on NTOs is a key element of the ERA. Its quality significantly depends on the careful identification of potentially exposed NTOs and meaningful testing in different containment steps. Concepts have been developed on how to select organisms from the receiving environments which would not dramatically increase the complexity of the ERA but significantly increase the scientific rigour and robustness of the results obtained[41,49]. These concepts comprise an exposure assessment, the selection of functional groups and the classification according to their biological functions. The goal is to focus on species which have an important functional role in a particular cropping system. Impacts on those species by GMP cultivation would actually result in an adverse environmental effect. The selected organisms should be used in toxicological tests in the laboratory, in the greenhouse or in field studies using the whole plant [50-52].

Protected Species Are Left Unconsidered In the ERA

Species, which are EU-wide, nationally or regionally endangered and/or protected may be of particular concern if GMPs interfere with their life cycles, habitats, competitors or food resources. Directive 2001/18/EC and its guidelines [15] specifically address endangered species.

However, protected species have not been seriously considered in any of the GM maize applications. This fact, however, has implications for the evaluation of the interactions of the GMP with NTOs. The only substitute species mentioned in *B. thuringiensis* maize applications is the Monarch butterfly, a species with virtually no relevance for European ecosystems. Even for maize 59122, which expresses a coleopteran-active protein, no non-pest, endangered or protected coleopteran species was considered.

We propose that the selected species for NTO testing should include key species which allow the evaluation of possible consequences for species of conservation concern. This may include species that are biologically or taxonomically close to endangered or protected species for which a thorough exposure assessment is necessary. Due to their vulnerability, highly protected species should not undergo toxicity testing. The lowest NOEC for sublethal effects for species ecologically or taxonomically close to endangered species should be used in the risk evaluation. Because the protection goal for those species differs compared with other NTOs, we strongly advise using assessment factors different from those used to assess overall biodiversity effects. The application of safety margins on the NOEC will then give an indication of the risk for a particular protected species.

However, the distribution of endangered or threatened organisms varies within Europe. For butterflies in Central Europe for instance, most threatened species can be found in eastern France, eastern Austria, the Romanian Carpathians and eastern Poland [53]. Threatened butterfly species as well as their protection status may also vary depending on the country [54]. So for the selection of relevant organisms, identification of regional differences in the protection status of particular species may be useful. Such species should be particularly considered during the ERA. Otherwise, ways must be found giving EU MS the necessary competence to address risks for endangered and protected species before GMOs are commercially placed on the market.

Field Assessments Do Not Take Place in Representative Environments

Field trials for agronomic evaluations of the GMPs took place overseas, in Europe or in both continents. The number of European countries (ranging from none to four, mostly two) and the number of locations within a specific country varied considerably between applications. Field studies for the assessment of potential effects on NTOs were more often conducted overseas than in Europe. Rarely was a specific assessment done in the same country and at the same location over two consecutive years. The choice of the locations was never justified by the applicants and the characterisation of the locations insufficient to judge their representativeness for European agronomic conditions.

We conclude that the analysed applications failed to adequately consider receiving environments relevant in the EU and therefore do not fulfil the requirements of Directive 2001/18/EC. The choice of field trial locations, however, is considered crucial in terms of representativeness for the agronomic and environmental conditions the GMP is expected to encounter when commercially cultivated. European environments differ substantially in biotic and abiotic conditions such as in the occurrence of TOs (e.g. *Sesamia* spp.) or NTOs, the number of generations of TOs (e.g. one or two generations of *O. nubilalis*), agricultural practices and agronomic structures (e.g. nitrogen input), cultivation systems (e.g. low-tillage farming), crop rotation practices or climatic conditions. These factors will influence adverse effect scenarios and as such are relevant for all three assessment categories analysed. Some effects may become evident only at a regional level. Toxin levels in the GMP and the infestation rates of target pests are influenced not only by soil management practices such as increased nitrogen levels [55-57] but also by other agricultural management practices such as planting time and irrigation [25,58]. In this respect, consideration of a sufficient number of growing seasons is important to take account of the variability of biotic (e.g. pest pressure) and abiotic (e.g. weather conditions) environmental factors.

Directive 2001/18/EC [1] and its guidance notes [15] address the necessity to provide data from different environments where the GMO will be used; hence, the potential receiving environment is emphasised (site-by-site or region-by-region principle) [15]. Since the authorisation of GMPs so far included all EU countries, it needs to reflect the full range of biogeographic regions. Hence, we consider it paramount that any GMP is tested for its performance, efficacy and potential adverse environmental effects under various and representative environmental conditions in Europe. This view is also in line with EFSA which requires that 'multiple geographical locations representative of the various environments in which the GMPs will be cultivated should be covered' [3]. In our view, the selection of representative environments in which a particular GMP will be tested has to be made at the beginning of the data collection of the ERA taking into account important agronomic and environmental factors. During this selection, considerations should be given to already existing concepts such as the agro-climatic zones for plant protection products [19], the European biogeographical regions mentioned in Directive 92/43/EEC [8] and the indicative map of European biogeographical regions developed for assistance to the Natura 2000 process [59].

No Consideration of Trait Interactions

Applications of GMPs derived from traditional crossing of two single-event GMPs (stacked event GMPs) generally referred to the assessments of the respective parental, single-event GMPs. Specific testing using the stacked event GM maize were hardly presented and potential interactions of transgene products when expressed in the same plant not addressed by the applicants. However, these interactions are relevant for the assessment of effects on and mediated by TOs as well as the evaluation of the interactions of the GMP with NTOs. Additionally, this approach ignores the legal requirements of the 'case-by-case' approach of Directive 2001/18/EC as well as scientific knowledge. For example, *B. thuringiensis* toxin interactions towards invertebrates, when bacteria-expressed but also when expressed in GMPs (e.g. in case of maize 59122), are well known, but constantly ignored in the current risk assessment practice of stacked GMPs (for a review, see[60]).

Stacked GM hybrids also need to undergo individual testing in the ERA. As for single-event GMPs, this testing should be guided by formulated research hypotheses and meaningful testing of the stacked GMP in different levels of containment to comply with the legislative requirements and the case-by-case principle of Directive 2001/18/EC.

CONCLUSIONS

From the evaluation of the data presented for the three assessment categories in GM maize applications, we conclude that there is a need for an improvement of the ERA and its data base if cultivation is envisaged. This requires a broader ERA concept that includes a thorough exposure analysis and the scientific evaluation of the whole GMP at different containment levels guided by testable risk hypotheses. Risk conclusions should be based on relevant data specifically generated for and using the GM event in question. Basic information on the agronomic characteristics and the efficacy of the GMP as well as its environmental behaviour must be provided for the ERA. The corresponding herbicide and its agronomic and environmental consequences also need to be taken into account. Such an approach would significantly strengthen the risk conclusions and their credibility.

Careful attention ought to be paid to the consideration of relevant data obtained from different European environments. Here, guidance for the selection of representative locations for field trials assessing the agronomic and environmental characteristics of the GMP is urgently needed. There is also a strong need for the development of standards with respect to the test designs, methods and parameters to achieve a common approach for specific assessments, e.g. the evaluation of the efficacy of the GMP against its target organism. Such standards could significantly improve the ERA without overruling the case-by-case approach.

The aforementioned shortcomings have led to considerable dispute among EU MS, in particular regarding the data gaps and uncertainties in the ERAs. Nevertheless, the presented data were frequently considered sufficient in EFSA GMO panel opinions and the GMP declared safe with respect to potential effects on the environment. This apparent discrepancy in the perception of environmental risks can only be resolved if a common understanding of how an ERA should be performed and which scientific data should be included will be achieved. In our view, the addressed suggestions would significantly improve the quality of the ERAs of GM maize applications and hence fulfil the requirements of Directive 2001/18/EC. EFSA has recently made an important step forward in updating the guidance on ERA in collaboration with all stakeholders [61]. This updated guidance is now under scrutiny by the MS.

AUTHORS' CONTRIBUTIONS

MD and MM drafted the manuscript with the help of MO. MD and MM analysed the data presented in the ERA of the GMP notifications for their relevance regarding the regulatory requirements. ME and AHi contributed to this part with a focus on the molecular characterisation and NTO, respectively. AHi, AHe, BT, HG, MD and MO conceived the study and participated in its the design. MD, MM and MO participated in the coordination of the study. All authors critically revised and contributed to the manuscript.

ACKNOWLEDGEMENTS

This study was supported by funding from the German Federal Environment Ministry (BMU) via the research and development project FKZ: 806 64 007 and is comprehensively documented in [14].

REFERENCES

1. European Commission: Directive 2001/18/EC of the European Parliament and of the Council of 12 March 2001 on the deliberate release into the environment of genetically modified organisms and repealing Council Directive 90/220/EEC - Commission Declaration.2001, 1-39. OJ L 106
2. Regulation (EC) No. 1829/2003 of the European Parliament and of the Council of 22 September 2003 on genetically modified food and feed 2003, 1-23. OJ L 268
3. European Food Safety Authority (EFSA): Guidance document of the Scientific Panel on Genetically Modified Organisms for the risk assessment of genetically modified plants and derived food and feed. *The EFSA Journal* 2006, 99:1-100.
4. Regulation (EC) No. 1830/2003 of the European Parliament and of the Council of 22 September 2003 concerning the traceability and labeling of genetically modified organisms and the traceability of food and feed products produced from genetically modified organisms and amending Directive 2001/18/EC 2003, 24-28. OJ L 268

5. European Commission: Communication from the Commission to the European Parliament, the Council, the Economic and Social Committee and the Committee of the Regions on the Freedom for Member States to decide on the cultivation of genetically modified crops. COM (2010)380 final
6. Andow DA, Hilbeck A: Science-based risk assessment for non-target effects of transgenic crops. *Bio-Science* 2004, 54:637-649.
7. Hilbeck A, Meier M, Römbke J, Jänsch S, Teichmann H, Tappeser B: Environmental risk assessment of genetically modified plants - concepts and controversies. *Environ Sci Eur* 2011, 23:13
8. Garcia-Alonso M, Jacobs E, Raybould A, Nickson T, Sowing P, Willekens H, van der Kouwe P, Layton R, Amijee F, Fuentes AM, Tencalla F: A tiered system for assessing the risk of genetically modified plants to non-target organism.*Environ Biosaf Res* 2006, 5:57-65.
9. Romeis J, Bartsch D, Bigler F, Candolfi MP, Gielkens MMC, Hellmich RL, Huesing JE, Jepson PC, Layton R, Quemada H, Raybould A, Rose RI, Schiemann J, Sears MK, Shelton AM, Sweet J, Vaituzis Z, Wolt JD: Assessment of risk of insect-resistant transgenic crops to non-target arthropods. *Nat Biotechnol* 2008, 26:203-208.
10. Report from the Commission to the Council and the European Parliament on the implementation of Regulation (EC) No. 1829/2003 of the European Parliament and of the Council on genetically modified food and feed COM (2006) 626 final
11. Second Report from the Commission to the Council and the European Parliament on the experience of Member States with GMOs placed on the market under Directive 2001/18/EC on the deliberate release into the environment of genetically modified organisms COM (2007) 81 final
12. Council of the European Union: Genetically modified organisms (GMOs) - Council conclusions of 5 December 2008. 16882/08.
13. Evaluation of the EU Legislative framework in the field of cultivation of GMOs under Directive 2011/18/EC and Regulation (EC) No. 1829/2003: Interim Report. DG Environment. A report submitted by EPEC. November 2009
14. Dolezel M, Miklau M, Hilbeck A, Eckerstorfer M, Heissenberger A, Gaugitsch H:Standardizing the environmental risk assessment of

genetically modified plants in the EU. [http://www.bfn.de/0301_veroe.html] *BfN Skripten 259* Bundesamt für Naturschutz, Bonn, Germany; 2009, 299.

15. European Commission: Commission Decision of 24 July 2002 establishing guidance notes supplementing Annex II to Directive 2001/18/EC of the European Parliament and of the Council on the deliberate release into the environment of genetically modified organisms and repealing Council Directive 90/220/EEC. *Official Journal of the European Communities* L 200/22. 2002/623/EC
16. Wolt JD, Conlan CA: Non-target exposure and risk assessment for cultivation of 1507 maize in Europe. *Dow AgroSciences LLC Study ID: GH-C 5214* 2001.
17. Bundessortenamt (BSA): Richtlinien zur Durchführung von landwirtschaftlichen Wertprüfungen und Sortenversuchen. [http://www.bundessortenamt.de/ internet30/ fileadmin/ Files/ PDF/ Richtlinie_LW2000.pdf] *Bundessortenamt* 2000.
18. European Commission: Directive 92/43/EEC: Council Directive of May 21st 1992 on the conservation of natural habitats and wild fauna and flora. 1996, 63. OJ L 59
19. Österreichische Agentur für Gesundheit und Ernährungssicherheit GmbH (AGES): Sorten und Saatgutblatt 2002. Methoden für Saatgut und Sorten gemäß **§5** Abs. 1 Saatgutgesetz 1997, BGBl. I Nr. 72 zgd. BGBl. I Nr. 109/2001. Richtlinien für die Sortenprüfung. *Schriftenreihe 59/2002 des BFL, 10. Jahrgang, Sondernummer 13*
20. European and Mediterranean Plant Protection Organization (EPPO/OEPP): Standards on efficacy evaluation of plant protection products issued by EPPO/OEPP: Nr. PP1/13(3) Ostrinia nubilalis; Nr. PP1/214(1): Principles of acceptable efficacy; Nr. PP1/223(1): Introduction to the efficacy evaluation of plant protection products; Nr. PP 1/241: Guidance on comparable climates; Nr. PP1/152(3): Design and analysis of efficacy evaluation trials.
21. Püntener W: Manual für Feldversuche im Pflanzenschutz. *Documenta CIBA-GEIGY AG Basel* 1981, 145.
22. Demaagd R, Weemen-Hendriks M, Molthoff JW, Naimov S: Activity of wild-type and hybrid *Bacillus thuringiensis* d-endotoxins against *Agrotis ipsilon*. *Arch Microbiol* 2003, 179:367.

23. Binning RR, Rice ME: Effects of transgenic *Bt* corn on growth and development of the stalk borer *Papaipema nebris* (Lepidoptera: Noctuidae). *J Econ Entomol* 2002, 95:622-627.

24. Archer TL, Schuster G, Patrick C, Cronholm G, Bynum ED, Morrison WP: Whorl and stalk damage by European and southwestern corn borers to four events of *Bacillus thuringiensis* transgenic maize. *Crop Prot* 2000, 19:181-190.

25. Archer TL, Patrick C, Schuster G, Cronholm G, Bynum ED, Morrison WP: Ear and shank damage by corn borers and corn earworms to four events of *Bacillus thuringiensis*transgenic maize. *Crop Prot* 2001, 20:139-144.

26. Horner TA, Dively GP, Herbert DA: Development, survival and fitness performance of*Helicoverpa zea* (Lepidoptera: Noctuidae) in MON810 *Bt* field corn. *J Econ Entomol* 2003, 96:914-924.

27. European Commission: Directive 91/414/EEC: Council Directive of 15 July 1991 concerning the placing of plant protection products on the market.1991, 1-32. OJ L 230

28. Rosati A, Bogani P, Santarlasci A, Buiatti M: Characterisation of 3′ transgene insertion site and derived mRNAs in MON810 YieldGard® maize. *Plant Mol Biol* 2008, 67:271-281.

29. Nguyen HT, Jehle JA: Quantitative analysis of the seasonal and tissue-specific expression of Cry1Ab in transgenic maize MON810. *J Plant Dis Prot* 2007, 114:82-87.

30. Brooks DR, Bohan DA, Champion GT, Haughton AJ, Hawes C, Heard MS, Clark SJ, Dewar AM, Firbank LG, Perry JN, Rothery P, Scott RJ, Woiwod IP, Birchall C, Skellern MP, Walker JH, Baker P, Bell D, Browne EL, Dewar AJ, Fairfax CM, Garner BH, Haylock LA, Horne SL, Hulmes SE, Mason NS, Norton LR, Nuttall P, Randle Z, Rossall MJ, Sands RJ, Singer EJ, Walker MJ: Invertebrate responses to the management of genetically modified herbicide-tolerant and conventional spring crops. I. Soil-surface-active invertebrates. *Philos Trans R Soc Lond B Biol Sci* 2003, 358:1847-1862. PubMed Abstract |Publisher Full Text | PubMed Central Full Text

31. Haughton AJ, Champion GT, Hawes C, Heard MS, Brooks DR, Bohan DA, Clark SJ, Dewar AM, Firbank LG, Osborne JL, Perry JN, Rothery P, Roy DB, Scott RJ, Woiwod IP, Birchall C, Skellern MP, Walker JH, Baker P, Browne EL, Dewar AJ, Garner BH, Haylock

LA, Horne SL, Mason NS, Sands RJ, Walker MJ: Invertebrate responses to the management of genetically modified herbicide-tolerant and conventional spring crops. II. Within-field epigeal and aerial arthropods. *Philos Trans R Soc Lond B Biol Sci* 2003, 358:1863-1877. PubMed Abstract |Publisher Full Text | PubMed Central Full Text

32. Hawes C, Haughton AJ, Osborne JL, Roy DB, Clark SJ, Perry JN, Rothery P, Bohan DA, Brooks DR, Champion GT, Dewar AM, Heard MS, Woiwod IP, Daniels RE, Young MW, Parish AM, Scott RJ, Firbank LG, Squire GR: Responses of plants and invertebrate trophic groups to contrasting herbicide regimes in the Farm Scale Evaluations of genetically modified herbicide-tolerant crops. *Philos Trans R Soc Lond B Biol Sci* 2003, 358:1899-1913.
33. Roy DB, Bohan DA, Haughton AJ, Hill MO, Osborne JL, Clark SJ, Perry JN, Rothery P, Scott RJ, Brooks DR, Champion GT, Hawes C, Heard MS, Firbank LG: Invertebrates and vegetation of field margins adjacent to crops subject to contrasting herbicide regimes in the Farm Scale Evaluations of genetically modified herbicide-tolerant crops. *Philos Trans R Soc Lond B Biol Sci* 2003, 358:1879-1898.
34. Krebs JR, Wilson JD, Bradbury RB, Siriwardena GM: The second Silent Spring? *Nature* 1999, 400:611-612.
35. Chamberlain DE, Fuller RJ, Bunce RGH, Duckworth JC, Shrubb M: Changes in the abundance of farmland birds in relation to the timing of agricultural intensification in England and Wales. *J Appl Ecol* 2000, 37:771-788.
36. Robinson RA, Sutherland WJ: Post-war changes in arable farming and biodiversity in Great Britain. *J Appl Ecol* 2002, 39:157-176.
37. Butler SJ, Vickery JA, Norris K: Farmland biodiversity and the footprint of agriculture. *Science* 2007, 315:381-383.
38. Birch ANE, Wheatley R, Anyango B, Arpaia S, Capalbo D, Getu Degaga E, Fontes E, Kalama P, Lelmen E, Lövei GL, Melo IS, Muyekho F, Ngi-Song A, Ochiendo D, Ogwang J, Pitelli R, Sétamou M, Sithanantham S, Smith J, Van Von N, Songa J, Sujii E, Quang Tan T, Wan F-H, Hilbeck A: Biodiversity and non-target impacts: a case study of *Bt*-maize in Kenya. In*Environmental Risk Assessment of Genetically Modified Organisms. A Case Study of*

Bt Maize in Kenya. Volume 1. Edited by Hilbeck A, Andow D. Wallingford: CAB International; 2004::117-185.

39. Andow DA, Lövei GL, Arpaia S: Ecological risk assessment for *Bt* crops. *Nat Biotechnol* 2006, 24:749-751.
40. Lang A, Lauber E, Darvas B: Early-tier tests insufficient for GMO risk assessment. *Nat Biotechnol* 2007, 25:35-36.
41. Andow DA, Zwahlen C: Assessing environmental risks of transgenic plants. *Ecol Lett* 2006, 9:196-214.
42. Hilbeck A, Andow DA, Arpaia S, Birch ANE, Chen Y, Fontes E, Fitt G, Lang A, Thi Thu Hong Le, Lövei GL, Manachini B, Nguyen TTC, Nguyen VH, Nguyen VT, Pham Van Lam, Van Toan Pham PC, Sujii E, Khuong Lai Trac UE, Wheatley R, Wilson L, Zwahlen C: Non-target and biological diversity risk assessment. In *Environmental Risk Assessment of Transgenic Organisms. Volume 4: Challenges and Opportunities with Bt Cotton in Vietnam*. Edited by Andow DA, Hilbeck A, Nguyen VT. Wallingford: CABI Publishing; 2008.
43. Herman RA, Scherer PN, Young DL, Mihaliak CA, Meade T, Woodsworth AT, Stockhoff BA, Narva KE: Binary insecticidal crystal protein from *Bacillus thuringiensis*, strain PS149B1: effects of individual protein components and mixtures in laboratory bioassays. *J Econ Entomol* 2002, 95:635-639.
44. Baumgarte S, Tebbe CC: Field studies on the environmental fate of the Cry1Ab *Bt*-toxin produced by transgenic maize (MON810) and its effect on bacterial communities in the maize rhizosphere. *Mol Ecol* 2005, 14:2539-2551.
45. Wandeler H, Bahylova J, Nentwig W: Consumption of two *Bt* and six non-*Bt* corn varieties by the woodlouse *Porcellio scaber*. *Basic Appl Ecol* 2002, 3:357-365.
46. Rosi-Marshall EJ, Tank JL, Royer TV, Whiles MR, Evans-White M, Chambers C, Griffiths NA, Pokelsek J, Stephen ML: Toxins in transgenic crop byproducts may affect headwater steam ecosystems. *PNAS* 2007, 104:16204-16208. PubMed Abstract | Publisher Full Text |PubMed Central Full Text
47. Harwood JD, Wallin WG, Obrycki JJ: Uptake of *Bt* endotoxins by nontarget herbivores and higher order arthropod predators: molecular evidence from a transgenic corn agroecosystem. *Mol Ecol* 2005, 14:2815-2823.

48. Harwood JD, Samson RA, Obrycki JJ: Temporal detection of Cry1Ab-endotoxins in coccinellid predators from fields of *Bacillus thuringiensis* corn. *Bull Entomol Res* 2007, 97:643-648.

49. Hilbeck A, Meier M, Raps A: Review of non-target organisms and *Bt* plants. [http:/ / www.greenpeace.org/ raw/ content/ international/ press/ reports/ review-on-non-target-organisms. pdf] *Ecostrat Report, Commissioned by Greenpeace International, Amsterdam* 2000.

50. Hilbeck A, Jänsch S, Meier M, Römbke J: *Analysis and Validation of Present Ecotoxicological Test Methods and Strategies for the Risk Assessment of Genetically Modified Plants. BfN Skripten 236, 116 pp. + Appendices*. Bonn, Germany: Bundesamt für Naturschutz; 2008.

51. Andow DA, Hilbeck A, Van Tuat Nguyen, (Eds): *Environmental risk assessment of genetically modified organisms: challenges and opportunities with Bt cotton in Vietnam. Volume 4*. Wallingford: CAB International; 2008.

52. Hilbeck A, Nelson K, Andow DA, Underwood E: Problem formulation and options assessment (PFOA) to assess the ecological risks associated with a GM crop. In *Risk, Hazard, Damage - Specification of Criteria to Assess Environmental Impact of Genetically Modified Organisms. Volume 1*. Edited by Breckling B, Verhoeven R. Naturschutz und Vielfalt 1; 2004::131-143.

53. Van Swaay C, Cuttelod A, Collins S, Maes D, López Munguira M, Šašić M, Settele J, Verovnik R, Verstrael T, Warren M, Wiemers M, Wynhof I: *European Red List of Butterflies*. Luxembourg: Publications Office of the European Union; 2010.

54. Van Swaay CAM, Warren MS: *Red Data book of European butterflies (Rhopalocera)*. Nature and Environment, No. 99. Strasbourg: Council of Europe Publishing; 1999.

55. Hilbeck A, Fontes E, Andow DA: *Environmental Risk Assessment of Transgenic Organisms. Volume 2: A Case Study of Bt Cotton in Brazil*. Wallingford: CAB International; 2006.

56. Phelan PL, Mason JF, Stinner BR: Soil-fertility management and host preference by European corn borer, Ostrinia nubilalis (Hübner), on *Zea mays* L.: a comparison of organic and conventional chemical farming. *Agric Ecosyst Environ* 1995, 56:1-8.

57. Bruns HA, Abel CA: Nitrogen fertility effects on *Bt* delta-endotoxin and nitrogen concentrations of maize during early growth. *Agron J* 2003, 95:207-211.

58. Bruns HA, Abel CA: Effects of nitrogen fertility on *Bt* endotoxin levels in maize. *J Entomol Sci* 2007, 42:35-43.

59. European Topic Centre on Biological Diversity (ETC/BD): *Indicative Map of European Biogeographical Regions: Methodology and Development*. European Topic Centre on Biological Diversity under contract with the European Environment Agency, Paris; 2006.

60. Spök A, Dolezel M, Eckerstorfer M, Freigassner M, Gaugitsch H, Heissenberger A, Karner S, Klade M, Proksch M, Schneider L, Treiber F, Uhl M: *Assessment of toxic and ecotoxic properties of novel proteins in GMOs*. Bundesministerium für Gesundheit, Familie und Jugend, Forschungsberichte der Sektion IV, Band; 2008:189.

61. EFSA Panel on Genetically Modified Organisms (GMO): Guidance on the environmental risk assessment of genetically modified plants.[http://www.efsa.europa.eu/en/gmo/gmoguidance.htm] *EFSA J* 2010, 8(11):1879. [111 pp.]

Chapter 6

Recruitment and Functionality Traits as Bioindicators of Ecological Restoration Success in the Lurg Hills District, Victoria, Australia

Kristin Monie, Singarayer Florentine, and Grant Palmer

Centre for Environmental Management, School of Science, Information Technology and Engineering, University of Ballarat, PO Box 663, Ballarat, Victoria, Australia

ABSTRACT

Introduction

The Regent Honeyeater Project commenced ecological restoration in the Lurg district in 1994, with an aim to restore habitats for the critically endangered Regent Honeyeater Anthochaera phrygiaand a range of other threatened and declining species. Within this context, our study aimed to explore whether plant reproduction can be an effective measure of ecological restoration success.

Methods

Evaluation involved comparing attributes at unrestored, restored and remnant sites to establish whether sites displayed evidence of a clear restoration trajectory. Five age classes (unrestored, 4–6 years old, 8–10 years old, 12–14 years old and remnant areas) and two landforms (upper hills and lower hills) were considered. The diversity of woody plant species—which have easily recognisable reproductive material and which all recruit seedlings—provides easily measured parameters that have the potential to allow the determination of early establishment success and long-term ecological development of restored ecosystems.

Results

Restoration plantings developed in a hybrid state towards a benchmark in the lower hills; seedling species composition differed significantly among age classes, increasing in similarity with time since restoration, with some divergence from the target pathway. Composition of functional groups with reproductive outputs was also significantly different among age classes; however, a restoration trajectory was only evident in the upper hills where sites converged towards the target goal.

Conclusions

Divergence or deviation from the restoration trajectory was not deemed to be a restoration failure, as the variety of functional groups with fruits and diversity of seedlings recruiting indicated a potential increase in resilience in the future due to greater variability across the landscape. Plant recruitment was effective in detecting development trends towards a restoration target in this study and therefore may be a useful measure that contributes to determining ecological restoration success.

INTRODUCTION

Monitoring and evaluation of restoration activities are critical steps toward developing and improving upon ecological restoration practices (Lindenmayer and Likens 2010; Lindenmayer et al.2012). However, despite vastly greater investments in on-ground works (Wilkins et al. 2003), the proportionally little monitoring of revegetated sites that has been conducted has sometimes been ineffectual in accurately determining ecological trajectories (Halle and Fattorini 2004). In this respect, the persistence of potentially ambiguous definitions of restoration activities and success (by practitioners), conflicting ecological theories (by academics) and a lack of clear guidelines (by regulators) has created some uncertainty about which parameters are the most important for determining the nature of ecological resilience and long-term stability. Whilst it is well documented that monitoring multiple variables is most valuable in determining restoration success (Ruiz-Jaen and Aide 2005a, b; Cristescu et al. 2013), resource limitations make it essential to identify a concise suite of attributes that can be confidently used as measures to rapidly demonstrate restoration outcomes—especially for large-scale projects (Menz et al. 2013). Another more contentious concern (addressed as a central theme of this special issue by Perring, Audet and Lamb, eds.) involves determining whether targeting the reinstatement of pre-disturbance reference ecosystems is always the most appropriate outcome and whether the post-disturbance ecosystems are on-course toward 'natural' (historic), 'hybrid' or 'novel' outcomes (Hobbs et al.2009). From a decidedly more pragmatic perspective, these factors underpin the challenges faced by practitioners, academics and regulators attempting to determine how similar field sampling sites should be prior to being deemed veritably or otherwise restored and, indeed, whether these sites are natural or novel.

So far, it appears that comparing sites with similar management histories across different points in time may more effectively establish whether sites are developing towards the restoration target or whether intervention is required to overcome specific ecological thresholds (Suding 2011). As is often the case for restoration sites in Australia (as elsewhere), which typically involve direct seeding practices as part of site preparation, vegetation composition and structure may change along natural succession pathways through the emergence and recruitment of seedlings (Clarke and Davison 2001). Populations may ultimately decline if recruitment rates are less than mortality rates (Clarke and Davison 2004), which makes this ratio a useful monitoring parameter among rehabilitated sites. The diversity of woody plant species—which have easily recognisable reproductive material and which all recruit seedlings—provides easily measured parameters that have the potential to allow the determination of early establishment success and long-term ecological development of restored ecosystems.

This project aimed to determine the effectiveness of monitoring such plant reproductive parameters in the Lurg Hills district, Victoria (Australia), as indicators of plant recruitment. Our study design compared the relative similarities between remnant, unrestored and different aged restoration sites with a particular investigative focus on (i) species composition of woody seedlings and (ii) composition of woody functional groups with buds, flowers or fruits present. Besides the obvious intention of determining restoration success of the sites themselves, this study also seeks to determine the utility of using plant recruitment and functionality traits as bioindicators for similar monitoring protocols.

Bioregional Description

The Lurg Hills district (36° 35′ S, 146° 07′ E) is located approximately 220 km north-east of Melbourne, Victoria, Australia (Figure 1). The region forms a connection between remnants of native vegetation occurring in the foothills of the Great Dividing Range and the Warby Ranges State Park (Thomas 2009). This landscape—which was largely cleared for agricultural purposes—receives approximately 668 mm mean annual rainfall, with mean temperatures ranging between 3.2 and 12.8°C in July and 14.9 and 31.0°C in January (Bureau of Meteorology 2011). During the given restoration study time frame (from 1996 to 2009), annual rainfall varied substantially (1996–2010), ranging from 243 mm in 2006 to 833 mm in 1996 at an average of 601 mm annually. Rainfall in the year the survey commenced was well above average at 946 mm following 4 notable years of below average rainfall.

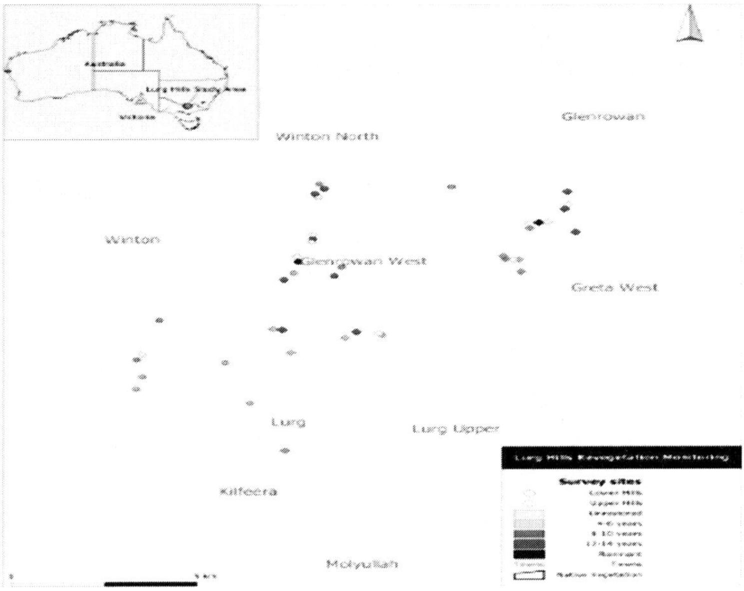

Figure 1: Location of study site and restoration sites.

The Regent Honeyeater Project (Thomas 2009) commenced ecological restoration in the Lurg district in 1994, with an aim to restore habitats for the critically endangered Regent HoneyeaterAnthochaera phrygia and a range of other threatened and declining species. Since 1996, restoration activities have occurred at over 490 sites, with 70 direct seeding sites and an additional 371 fenced sites (to restrict livestock grazing), to protect over 1,378 ha of habitat (Thomas 2012). Restoration sites generally differed in their condition prior to treatment, but all were previously grazed and/or cropped. Targets for restoration were sites that (i) adjoined remnant vegetation, (ii) incorporated isolated paddock trees and (iii) created connections between vegetated areas by enhancing or widening roadside vegetation and creek lines (Thomas 2009). Tubestock planting was used for the majority of restoration, with direct local provenance seeding also used increasingly over the years. Sites were generally prepared by ripping dry soils, followed by herbicide application after germination of winter weeds, and were ultimately fenced-in to exclude livestock (Thomas 2009). Between 35 and 40 tree, shrub and groundcover species were planted manually after scalping the upper surface to remove the weed seed, and carton guards were used to protect plants. Follow-up weed control efforts were not consistent across all restoration sites, and replacement plantings were undertaken in following years at sites with high losses, which were due to damage from wildlife. No artificial irrigation methods were used with the exception of sites planted very late in the season.

METHODS

Sampling Design

A total of 438 sites were restored between 1996 and 2010 (Thomas 2012). Factorial blocking design and elimination criteria were used to account for natural variation and other differences among restoration sites. Elimination criteria included size, fencing and restoration method. The sites were between 1 and 7 ha in size, continuously fenced from stock since restoration, and planted with tubestock only. Two blocking factors of interest were used to further minimise confounding variables, these being (i) restoration age and (ii) landform and vegetation community characteristics.

Restoration Age

Sites were categorised according to the time since the restoration planting. Space-for-time substitution was used to capture the different number of years since restoration. Three periods were selected to reflect significantly different stages of development and growth, with sites less than 4 years since planting being excluded from selection, as some species may not have reached reproductive maturity. This was in line with Wilkins et al. (2003), who found only slight differences in floristic composition between unrestored and restored sites up to 9 years old. The restoration ages selected for evaluation were: 4–6 years old (planted in 2004–2006); 8–10 years (2000–2002); and 12–14 years (1996–1998).

Landform and Vegetation

Lurg landform categories were created using distinguishing landform characteristics detailed in Speight (2009), in combination with descriptions in the Revegetation Guide for the Goulburn Broken Catchment (Earl et al. 2001). Elevations were measured against the lowest location in the district, which was the Winton Wetlands, at approximately 170 m above sea level. Sites were allocated to landform categories based on mapped contour lines, and those sites on the boundary between two categories were allocated subjectively to the appropriate category using this mapping as a guide.

Landform categories were compared against modelled ecological vegetation classes [ecological vegetation classes (EVCs) are vegetation community classifications based on a combination of floristic, structure (life forms) and ecological characteristics and landscape position (Department of Sustainability and Environment 2013)] in the study area. Two broad landform categories (upper hills, lower hills) were selected from which the study sites were chosen, with the remaining categories (granitic hills, plains and wetlands) excluded due to inadequate replicates across the chosen age groups. The upper hills incorporated the dry steeper slopes of eucalypt-dominated forest with a shrubby or grassy understorey (Figure 2). The lower hills were gentler slopes of eucalypt woodland, with a sparse understorey of shrubs, herbs and grasses and with leaf litter dominating the groundlayer (Figure 3).

Figure 2: Vegetation community of the upper hills: eucalypt forest with a shrubby or grassy understorey (EVC shrubby dry forest / grassy dry forest).

Figure 3: Vegetation community of the lower hills: eucalypt woodland with a sparse understorey of shrubs, herbs and grasses and with leaf litter dominating the groundlayer (EVC box ironbark forest).

Site Selection

A total of 42 sites were sampled equally across the two landforms (Table 1), each of which contained 15 restored sites (comprising 5 in each of three age categories), five unrestored sites and one remnant site, with the restored sites being randomly selected to ensure site selection was objective and unbiased. The unrestored (pre-restoration) sites facilitated comparison between restored and unrestored sites, allowing a determination of the effects of restoration. Due to the small sample size and the large number of potential unrestored sites under a range of management regimes, these sites were purposefully selected, based on management regimes, to represent the range of conditions. In addition, intact reference sites were difficult to locate due to the history of clearing in the district, thus only one remnant site was sampled in each landform category.

Table 1: Number of sites sampled in each restoration age class and landform combination

Restoration age class	Lower hills	Upper hills
Unrestored (pasture)	5	5
Restored (4–6 years old)	5	5
Restored (8–10 years old)	5	5
Restored (12–14 years old)	5	5
Remnant vegetation	1	1

Monie et al.

Monie et al. Ecological Processes 2013 **2**:27 doi:10.1186/2192-1709-2-27

Data Collection and Analysis

Data were collected from 337 quadrats across 42 sites in the summer of 2010–2011. The number of quadrats sampled at a site differed to ensure that the area surveyed was proportional to the size of each site. Dimensions of the 120 m^2 quadrats differed based on the shape of the restoration site. A systematic search was conducted to locate planted and regenerating individuals in each quadrat. The species type and the height and diameter at base were recorded for each plant. Diameters were measured using Vernier calipers or a diameter measuring tape where the diameter exceeded 150 mm. Heights were measured using a 2 m ruler or a Silva clinometer for taller plants. The presence of reproductive material (buds, flowers, fruits) was recorded for each plant. Remnant trees were measured using the same method if they were present within a sampled quadrat. Recruitment was considered episodic if there were large numbers of individuals of the same age class and species within a quadrat (for example, if there were river red-gum seedlings in a drainage line). Where regeneration was clearly arising from suckers, this was noted due to the different reproductive method.

Multivariate analysis, using Primer 6© (Primer-E, Ltd, Lutton, UK), was conducted to compare similarities between sites of different age classes using analysis of similarity (ANOSIM) and Bray-Curtis dissimilarity measures. Similarity percentage analysis (SIMPER) was then conducted to investigate which variables contributed most to these differences. Within-group similarities were not calculated for the remnant due to the absence of replicates. Multidimensional scaling (MDS) plots, using Bray-Curtis dissimilarity measures, were used to create an ordination of the site data. Data at each site were pooled and the mean number of seedlings per quadrat determined. Functional groups were defined based on lifeforms in the Habitat Hectares approach (Department of Sustainability and Environment 2004). Seedlings were allocated to groups based on description and height.

RESULTS

Seedling Species Composition

Seedlings of various species were recorded in 88% of sites. There was evidence of recruitment at most of the restored and remnant sites in the upper hills, the exceptions being two restored sites (4–6 and 12–14 years old). In contrast, there was virtually no evidence of recruitment at unrestored sites in the lower hills, where only one seedling was recorded. Thirty-seven species of trees and shrubs (from eight families) were regenerating across the surveyed sites, 22 of which occurred in both landform categories. Eucalypts were mainly found regenerating near large old remnant trees. A rare species in the district, the western golden-tip Goodia medicaginea, was recruited in small numbers at both landforms. The total number of species recruiting differed by age class, ranging from 0 to 18 in the lower hills, and 0 to 20 in the upper hills. No woody weeds were recorded in any sites.

Lower Hills

Seedling species composition differed significantly among age classes in the lower hills (global $R = 0.378$, $p = 0.001$). ANOSIM pairwise comparisons indicated the age classes that were most different in seedling species composition were unrestored and restored (12–14 years) (Table 2). Other age classes that differed were unrestored and restored (8–10 years); restored (4–6 years) and restored (8–10 years); and restored (8–10 years) and restored (12–14 years). The MDS ordination (Figure 4) showed older restored sites were closer in seedling composition to the remnant than the younger age classes were. Variation between sites of the same age decreased with time since restoration, with most variation occurring between restored (4–6 years) sites. This is supported by the SIMPER analysis, which showed within-group similarities increased with restoration age from 8.21% similarity in restored (4–6 years), to 19.23% in restored (12–14 years). These similarities were due to a few species: Acacia dealbata (51.28%) in restored (4–6 years), Dodonaea viscosa ssp. angustissima (76.23%) in restored (8–10 years) and Acacia pycnantha (76.61%) in restored (12–14 years). Dissimilarity between age classes were due to a range of species Species which contributed most to within-group similarity were also largely responsible for between-group dissimilarity. Sites became more like the remnant with increased time since restoration.

Table 2: Species composition of seedlings in lower hills

Age class	Unrestored		Restored (4–6)		Restored (8–10)		Restored (12–14)	
	R statistic	p value	R statistic	p value	R statistic	p value	R statistic	p value
Restored (4–6)	0.063	0.183						
Restored (8–10)	0.571	0.008	0.364	0.008				
Restored (12–14)	0.697	0.008	0.210	0.087	0.512	0.008		
Remnant	0.000	1.000	0.040	0.333	0.240	0.500	0.120	0.667

Species composition of seedlings in different aged sites in the lower hills (ANOSIM pairwise comparisons showing R statistic and significance level, with significant p-values in bold. Global R = 0.378, p = 0.001).

Monie et al.

Monie et al. Ecological Processes 2013 2:27 doi:10.1186/2192-1709-2-27

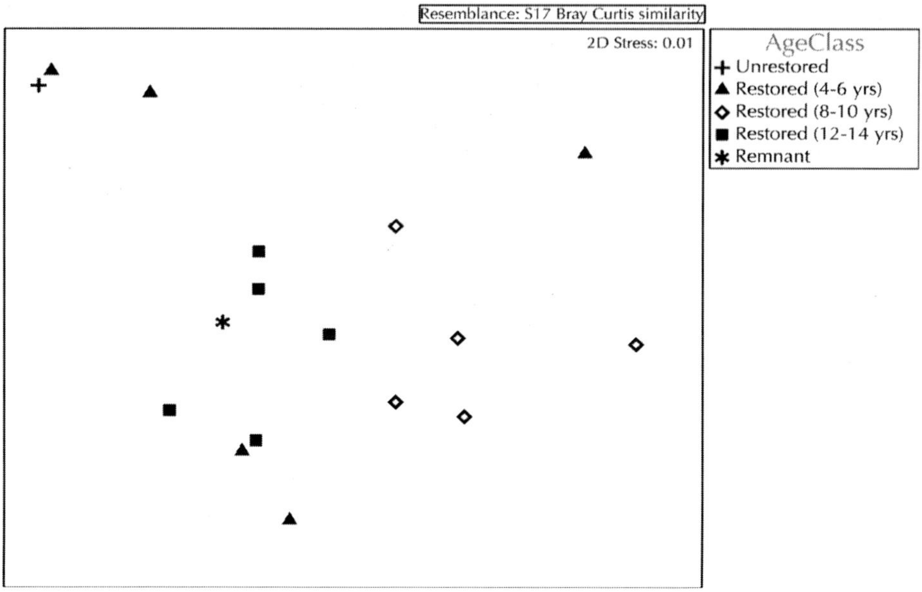

Figure 4: Species composition of seedlings in lower hills. MDS ordination of species composition of seedlings in different age classes in lower hills, with subset of data enlarged (four unrestored sites excluded from frame as clearly separated from remaining cluster).

Upper Hills

No significant differences were found in the species composition of seedlings between different age classes in the upper hills (ANOSIM global R statistic 0.114; $p = 0.068$).

REPRODUCTIVE OUTPUTS

Buds and Flowers

Lower Hills

The composition of functional groups with buds or flowers was significantly different among age classes ($R = 0.254$, $p = 0.003$).

ANOSIM pairwise comparisons revealed the differences occurred between the unrestored and both restored (8–10 years) and restored (12–14 years) age classes (Table 3). SIMPER analysis showed that the main sources of variability between unrestored sites and the restored (8–10 years) and restored (12–14 years) age classes were immature trees and medium shrubs, and immature trees respectively. Unrestored sites generally decreased in dissimilarity to the unrestored age class with time since restoration. Dissimilarities between the remnant and all other groups were alike, which was mainly due to the higher number of canopy trees with buds or flowers in the remnant. Unrestored and the youngest restored sites had low within-group similarities (20.37 and 15.68% respectively); this increased with restoration age (46.65 and 51.58% similarity for 8–10 years and 12–14 years respectively). Younger sites had a more diverse mix of small and medium shrubs, immature trees and trees with flowers.

Table 3: Composition of functional groups with buds or flowers in lower hills

Age class	Unrestored		Restored (4–6)		Restored (8–10)		Restored (12–14)	
	R statistic	p value	R statistic	p value	R statistic	p value	R statistic	p value
Restored (4–6)	0.233	0.095						
Restored (8–10)	0.467	0.008	0.040	0.317				
Restored (12–14)	0.516	0.008	0.180	0.103	0.006	0.413		
Remnant	-0.111	0.500	-0.180	0.667	0.680	0.167	0.760	0.167

Composition of functional groups with buds or flowers in different aged sites in the lower hills (ANOSIM pairwise comparisons showing R statistic and significance level, with significant p-values in bold. Global R = 0.254, p = 0.003).

Monie et al.

Monie et al. Ecological Processes 2013 **2**:27 doi:10.1186/2192-1709-2-27

Upper Hills

The composition of functional groups with buds or flowers was significantly different among age classes (R = 0.543; p = 0.001). ANOSIM pairwise comparisons (Table 4) indicated the differences occurred between unrestored and all restored age classes, and between restored

(4–6 years) and restored (12–14 years). SIMPER analysis showed that differences between unrestored and restored sites were mostly due to medium shrubs in the two younger restored age classes, and immature trees in the two older restored age classes. Medium shrubs at restored (4–6 years) sites and immature trees at restored (12–14 years) sites were largely responsible for differences between the two age classes. Dissimilarity decreased with time since restoration for all age classes. Unrestored sites in the upper hills contained no woody plants with buds or flowers. Restored age classes increased slightly in within-group similarity with age (52.99, 55.62 and 59.42%), mainly due to medium shrubs in the two younger restored groups and immature trees in the oldest restored age class.

Table 4: Composition of functional groups with buds or flowers in upper hills

Age class	Unrestored		Restored (4–6)		Restored (8–10)		Restored (12–14)	
	R statistic	p value	R statistic	p value	R statistic	p value	R statistic	p value
Restored (4–6)	0.714	0.008						
Restored (8–10)	0.703	0.008	0.248	0.087				
Restored (12–14)	0.703	0.008	0.658	0.008	0.228	0.056		
Remnant	-0.200	1.000	0.840	0.167	0.200	0.333	0.440	0.167

Composition of functional groups with buds or flowers in different aged sites in the upper hills (ANOSIM pairwise comparisons showing R statistic and significance level, with significant p-values in bold. Global R = 0.543, p = 0.001).

Monie et al.

Monie et al. Ecological Processes 2013 **2**:27 doi:10.1186/2192-1709-2-27

Fruits

Lower Hills

ANOSIM analysis demonstrated significant differences in the average number of woody plants with fruits in each functional group among age classes (R = 0.318; p = 0.001). These differences occurred between unrestored and all restored age classes (Table 5), and between restored (4–6 years) and restored (12–14 years). Unrestored sites differed from restored sites mainly due to medium shrubs with fruits in restored (4–6 years) sites, and both medium shrubs and immature trees in the two older restored age classes file 4: Appendix 4). Differences between restored (4–6 years) and restored (12–14 years) age classes were mainly due to the abundance of immature trees and medium shrubs with fruits. The unrestored age class was most similar to the remnant due to the presence of canopy trees and a lack of shrubs with fruits in both age classes. Restored age classes were most similar to sites closest in age.

Table 5: Composition of functional groups with fruits in lower hills

Age class	Unrestored		Restored (4–6)		Restored (8–10)		Restored (12–14)	
	R statistic	p value	R statistic	p value	R statistic	p value	R statistic	p value
Restored (4–6)	0.428	0.024						
Restored (8–10)	0.520	0.008	0.068	0.246				
Restored (12–14)	0.474	0.008	0.218	0.032	-0.052	0.620		
Remnant	-0.111	0.500	0.520	0.330	1.000	0.167	1.000	0.167

Composition of functional groups with fruits in different aged sites in the lower hills (ANOSIM pairwise comparisons showing R statistic and significance level, with significant p-values in bold. Global R = 0.318, p = 0.001).

Monie et al.

Monie et al. Ecological Processes 2013 **2**:27 doi:10.1186/2192-1709-2-27

Similarities within age classes differed: 20.37% (unrestored), 27.43% (restored 4–6 years), 51.72% (restored 8–10 years) and 48.48% (restored 12–14 years). Canopy trees were the main source of similarity in the unrestored age class, while medium shrubs and immature trees were the main source in the restored classes.

Upper Hills

Functional group composition with fruits differed significantly among age classes (R = 0.568, p = 0.001). All age classes were significantly different from each other except for the remnant (Table 6). Differences were mostly due to the abundance of medium shrubs with fruits (all restored age classes), and immature trees and trees at older restored sites Dissimilarities between unrestored and other age classes decreased with age, due to the minor presence of remnant canopy trees (with fruits) in the unrestored age class. Remnants had more in common with older restored sites (8–10 years and 12–14 years). Restored age classes were more similar to restored sites closest in age. Restored (4–6 years) sites were most similar (74.42%) to each other due to the presence of medium shrubs with fruits. Similarity was lower for the older restored age classes: restored (8–10 years) (59.17%), restored (12–14 years) (55.05%); however, the number of functional groups with fruits increased to include immature trees, plus trees in the oldest restored age class.

Table 6: Composition of functional groups with fruits in upper hills

Age class	Unrestored		Restored (4–6)		Restored (8–10)		Restored (12–14)	
	R statistic	p value	R statistic	p value	R statistic	p value	R statistic	p value
Restored (4–6)	0.647	0.008						
Restored (8–10)	0.619	0.008	0.460	0.016				
Restored (12–14)	0.619	0.008	0.856	0.008	0.260	0.040		
Remnant	-0.200	0.667	1.000	0.167	0.680	0.167	0.640	0.167

Composition of functional groups with fruits in different aged sites in the upper hills (ANOSIM pairwise comparisons showing R statistic and significance level, with significant p-values in bold. Global R = 0.568, p = 0.001).

Monie et al.

Monie et al. Ecological Processes 2013 **2**:27 doi:10.1186/2192-1709-2-27

DISCUSSION

Our study addresses rather typical restoration works in Australia (i.e., returning natural/native ecosystems following agricultural development), and some of the difficulties faced when attempting to define ecosystem development toward a given restoration target. In this regard, restoration plantings in the Lurg Hills generally developed in different ways across the two landforms studied (i.e., upper and lower hills). All sites were found in a hybrid state sharing some components with the remnant (historical) ecosystems, but also some novel proportional combinations of species assembly. Despite differences between groups, trends towards the restoration target were evident for species composition of seedlings only in the lower hills, and for bud/flower and fruit production only in the upper hills. For example, after 8 years, restored sites in the lower hills had a seedling composition that was very different to both unrestored and younger restored sites. Sites also became more similar to the remnant with time, particularly 12 years after restoration.

A range of development scenarios have been identified where sites may converge, diverge or deviate from the restoration target (Suding 2011). The majority of sites in the lower hills converged towards the target and also increased in similarity to sites of the same age class with time. This supports the idea that, as a community ages, the number of potential development trajectories may decrease due to plant competitive ability (Nuttle et al. 2004). However, a range of factors may have influenced these similarities within age classes. As Hobbs and Norton (2004) point out, the sites' starting conditions, order of species introductions and subsequent management will all interact to affect how a community will develop. Seed or microsite availability can also have a strong influence on the composition and structure of ecosystems through the failure of species to recruit (Clark et al. 2007). There were some instances where seedlings seemed to derive from adjoining remnant vegetation, as evidenced by patterns of seedling dispersal in relation to remnant trees, maturity of restored vegetation and restoration species lists. However, the majority of species recruiting were the same as those planted during restoration. While seeds were not tested for viability, it is likely much of the regeneration resulted from mature planted individuals (e.g., fruits were present on medium shrubs after only 4 years). Many studies show that recruitment of trees and shrubs is often affected to a greater extent by the availability of microsites, areas with suitable conditions for germination and growth (Clarke and Davison 2001; Clark et al. 2007; Gómez-Aparicio 2008). However, as the extent of potential recruitment is ultimately determined by the availability of viable seed (Clarke and Davison 2001), the production of adequate seed in restored ecosystems is clearly essential for achieving a target community. Introducing seed of additional species may help reduce any seed limitations (Hobbs and Norton2004; Young et al. 2005; Clark et al. 2007) which may be caused by decreased pollination and dispersal opportunities (Bennett et al. 2009).

The lack of recruitment at two restored sites (4–6 years and 12–14 years) in the lower hills suggests divergence, which occurs when sites that have undergone similar restoration develop along different trajectories (Suding 2011). Structurally, these sites differ from other sites of the same age, which have a range of species recruiting. It would be worth investigating the barriers that are preventing recruitment, such as herbivory, microsite variations or seed availability. Some species present in reference ecosystems may not re-establish under current conditions unless regeneration niches are provided (Zedler et al. 2012), yet recruitment of seedlings 8 years after restoration indicates niches have been provided for many indigenous species—including rare species such as Goodia medicaginea, which was found recruiting in small numbers in both landforms. In addition, no woody weeds were recorded at either landform. Therefore, seedlings and the structure of the tree and shrub layers were composed solely of remnants or planted indigenous species sourced from local provenance.

Zedler et al. (2012) argue that sites that need constant redirection towards their target ecosystem might be more sustainable if retained as novel ecosystems, thereby acknowledging the barriers preventing veritable restoration. With the exception of the two sites lacking seedlings, if diverse recruitment of indigenous species continues to occur, hybrid ecosystems could be retained in preference to novel ones, and have the potential to become more similar to a natural or historical ecosystem with time. If biota cannot regenerate, development into a novel ecosystem will probably occur (Hobbs et al. 2009). The insignificant difference in species composition of seedlings in the upper hills may be due to further factors (not investigated in the present context) involving more subtle or site-specific management practices, e.g., due to different historical land-usage intensities. Alternatively, potentially large variability in abiotic factors across the steeper upper hills (such as aspect, slope, moisture availability and proximity to larger patches of remnant vegetation) may be an influence.

Restored sites had buds, flowers and fruits after 4 years in the upper hills, with sites becoming more similar to the remnant as rapidly as 8 years after restoration. The absence of a clear trend towards a restoration target for bud/flower and fruit production in the lower hills may be due to the presence of canopy trees in both the unrestored and remnant sites, and the diverse range of functional groups with buds/ flowers or fruits in the restored sites, which were fewer in the remnant. This resulted in greater similarity between the remnant and unrestored sites. While the more fertile, gentler slopes of the lower hills may have influenced the production of buds/flowers and fruits on a diverse range of functional groups, the different stages of ecological development of the restored sites and the remnant may have been a factor. As only one remnant was surveyed in each landform, this was a limitation of the study. The different vegetation structure in the late stage of ecological development of the remnant is also likely to have an impact when assessing restoration success; as Suding and Hobbs (2009) point out, each reference site may reflect only one of the many stages of development, any of which would form the model for restoration.

Has Restoration in the Lurg Hills been Successful?

Despite much debate surrounding the use of remnant ecosystems as restoration targets (Hobbs and Harris 2001; Harris et al. 2006; Hobbs et al. 2009; Palmer et al. 2006; Choi 2007; Comín2010), these selected reference systems are still commonly used to determine site-specific restoration success. In this regard, seedling recruitment and reproductive parameters provide limited insight into whether sites are developing towards a restoration target. The establishment of self-sustaining ecosystems is an ecological attribute the Society for Ecological Restoration International Science and Policy Working Group (2004) proposes using to evaluate restoration success. Given seedlings recruited in most sites in the Lurg Hills, most of the restoration could be deemed successful at this point in time. The presence of seedlings signifies a life cycle has successfully been completed, as transplanted individuals have flowered, set fruit, dispersed seed and recruited new seedlings (Menges 2008). In addition, microsite conditions have clearly been conducive to germination and growth. While it may be preferable from a management perspective for sites to converge together along a predictable pathway towards a restoration target, Suding (2011) argues that resilience may be increased if there is variability at a landscape scale. Resilience is vital for an ecosystem to adapt and recover from disturbances, while still maintaining ecological function and health (Folke et al. 2004; Walker and Salt 2006; Clewell and Aronson2007). Diversity is thought to influence ecosystem resilience (Grant 2009; Allen et al. 2010); therefore, variability among sites of the same age may in fact be beneficial in the longer term, increasing resilience through differences in diversity as seedlings mature and alter vegetation structure and composition. This is particularly relevant under uncertain future conditions including climate change, as increased human demand places further pressure on ecosystems, which may not remain stable (Folke et al. 2004).

CONCLUSIONS

The majority of ecological restoration sites in the Lurg Hills have successfully recruited woody seedlings from a range of species, as early as 4 years after restoration. Most sites are in a hybrid state, developing towards the restoration target and becoming more similar to the remnant site and more distant from unrestored sites with age. However, this differed between landforms. The diversity of functional groups producing buds, flowers or fruits is also promising for the future of these restoration sites. Intervention may be useful at some sites where no recruitment occurred, by introducing additional seed or creating further opportunities for seed germination and establishment through targeted disturbance. Provided diversity is maintained and there are dispersal and recruitment opportunities through landscape connectivity, the deviation or divergence of some sites from the restoration target may not be an issue as resilience may potentially be increased by variability across the landscape. Plant recruitment was effective in detecting development trends towards a restoration target in this study and therefore is considered a useful measure that contributes to detecting ecological restoration success.

AUTHORS' CONTRIBUTIONS

KM gathered, analysed and interpreted the data and primarily drafted the manuscript. SF and GP provided initial research ideas, design and refined methodology. SF and GP provided editorial commentary. All authors read and approved the final manuscript.

ACKNOWLEDGEMENTS

Authors would like to thank Ray Thomas of the Regent Honeyeater Project for his willingness to provide access to project file notes and maps. We thank Dale Robinson, John Miller, Lucinda Monie, Kim Magnay, Tim D'Ombrain, Jacqui Monie and Belinda Taylor for their help in the field. We thank landholders in the Lurg district for allowing us to work on their properties. We also thank Tony Wilson and Tricia Wevill for statistical techniques. We thank Drs. Patrick Audet (The University of Queensland, Queensland) and Michael Perring (University of Western Australia, Perth) for providing constructive feedback on the earlier versions of our manuscript. The financial support was provided by The Regional Futures Grant, University of Ballarat.

REFERENCES

1. Allen CR, Gunderson LH, Holling CS (2010) Commentary on part one articles. In: Allen CR, Holling C, Gunderson LH (eds) Foundations of ecological resilience, Washington DC: Island Press.
2. Bennett AF, Haslem A, Cheal DC, Clarke MF, Jones RN, Koehn JD, Lake PS, Lumsden LF, Lunt ID, Mackey BG, Mac Nally R, Menkhorst PW, New TR, Newell GR, O'Hara T, Quinn GP, Radford JQ, Robinson D, Watson JEM, Yen AL (2009) Ecological processes: a key element in strategies for nature conservation. Ecol Manage Restor 10(3):192-199
3. Bureau of Meteorology (2011) Climate statistics for Australian locations: monthly climate statistics.http://www.bom.gov.au/climate/averages/tables/cw_082002.shtml Accessed 15 Dec 2011
4. Choi YD (2007) Restoration ecology to the future: a call for a new paradigm. Restor Ecol 15(2):351-353

5. Clark CJ, Poulsen JR, Levey DJ, Osenberg CW (2007) Are plant populations seed limited? A critique and meta-analysis of seed addition experiments. Am Nat 170(1):128-150
6. Clarke PJ, Davison EA (2001) Experiments on the mechanism of tree and shrub establishment in temperate grassy woodlands: seedling emergence. Aust Ecol 26:400-412
7. Clarke PJ, Davison EA (2004) Emergence and survival of herbaceous seedlings in temperate grassy woodlands: recruitment limitations and regeneration niche. Aust Ecol 29:320-331
8. Clewell AF, Aronson J (2007) Ecological restoration: principles, values, and structure of an emerging profession. Washington DC: Island Press.
9. Comín FA (2010) The challenges of humanity in the twenty-first century and the role of ecological restoration. In: Comín F (ed) Ecological restoration: a global challenge, New York: Cambridge University Press.
10. Cristescu RH, Rhodes J, Frére C, Banks PB (2013) Is restoring flora the same as restoring fauna? Lessons learned from koalas and mining rehabilitation. J App Ecol 50:423-431
11. Department of Sustainability and Environment (2004) Native vegetation: sustaining a living landscape. Vegetation quality assessment manual: guidelines for applying the habitat hectares scoring method. Version 1.3. Melbourne, Australia: Department of Sustainability and Environment.
12. Department of Sustainability and Environment (2013) Native vegetation – information. http://www.dse.vic.gov.au/land-management/land/native-vegetation-home/native-vegetation-information webcite. Accessed 23 Apr 2013
13. Earl G, Stelling F, Titcumb M, Berwick S (eds) (2001) Revegetation guide for the Goulburn broken catchment East Melbourne: Dept. of Natural Resources & Environment.
14. Folke C, Carpenter S, Walker BS, Elmqvist T, Gunderson L, Holling C (2004) Regime shifts, resilience, and biodiversity in ecosystem management. Annu Rev Ecol Evol Syst 35:557-581
15. Gómez-Aparicio L (2008) Spatial patterns of recruitment in Mediterranean plant species: linking the fate of seeds, seedlings

and saplings in heterogeneous landscapes at different scales. J Ecol 96:1128-1140

16. Grant C (2009) State-and-transition models for mining restoration in Australia. In: Hobbs R, Suding K (eds) New models for ecosystem dynamics and restoration, Washington DC: Island Press. pp 280-294
17. Halle S, Fattorini M (2004) Advances in restoration ecology: insights from aquatic and terrestrial ecosystems. In: Temperton V, Hobbs R, Nuttle T, Halle S (eds) Assembly rules and restoration ecology: bridging the gap between theory and practice, Washington DC: Island Press. pp 10-33
18. Harris JA, Hobbs RJ, Higgs E, Aronson J (2006) Ecological restoration and global climate change. Restor Ecol 14(2):170-176
19. Hobbs R, Harris J (2001) Restoration ecology: repairing the Earth's ecosystems in the new millenium. Restor Ecol 9(2):239-246
20. Hobbs RJ, Higgs E, Harris JA (2009) Novel ecosystems: implications for conservation and restoration. Trends Ecol Evol 24(11):599-605
21. Hobbs RJ, Norton DA (2004) Ecological filters, thresholds, and gradients in resistance to ecosystem reassembly. In: Temperton V, Hobbs R, Nuttle T, Halle S (eds) Assembly rules and restoration ecology: bridging the gap between theory and practice, Washington DC: Island Press. pp 72-95
22. Lindenmayer DB, Likens GE (2010) Effective ecological monitoring. Collingwood, VIC: CSIRO.
23. Lindenmayer DB, Likens GE, Anderson A, Bowman D, Bull CM, Burns E, Dickman CR, Hoffmann AA, Keith DA, Liddell MJ, Lowe AJ, Metcalfe DJ, Phinn SR, Russell-Smith J, Thurgate N, Wardle GM (2012) Value of long-term ecological studies. Aust Ecol 37:745-757
24. Menges ES (2008) Turner review no. 167. Restoration demography and genetics of plants: when is a translocation successful? Aust J Bot 56(3):187-196 Text
25. Menz MHM, Dixon KW, Hobbs RJ (2013) Hurdles and opportunities for landscape-scale restoration. Science 339:526-527

26. Nuttle T, Hobbs RJ, Temperton VM, Halle S (2004) Assembly rules and ecosystem restoration: where to from here? In: Temperton V, Hobbs R, Nuttle T, Halle S (eds) Assembly rules and restoration ecology, Washington DC: Island Press. pp 410-421
27. Palmer MA, Falk DA, Zedler JB (2006) Ecological theory and restoration ecology. In: Foundations of restoration ecology. Island Press, Washington DC, pp. pp 1-10
28. Ruiz-Jaen MC, Aide TM (2005) Restoration success: how is it being measured? Restor Ecol 13(3):569-577
29. Ruiz-Jaen MC, Aide TM (2005) Vegetation structure, species diversity, and ecosystem processes as measures of restoration success. For Ecol Manage 218(1–3):159-173
30. Society for Ecological Restoration International Science and Policy Working Group (2004) The SER international primer on ecological restoration. Washington DC: SER.
31. Speight JG (2009) Landform. In: National Committee on Soil and Terrain (ed) Australian soil and land survey field handbook, 3rd edn. Melbourne: CSIRO Publishing.
32. Suding KN (2011) Toward an era of restoration in ecology: successes, failures, and opportunities ahead. Annu Rev Ecol Evol Syst 42:465-487
33. Suding KN, Hobbs RJ (2009) Models of ecosystem dynamics as frameworks for restoration ecology. In: Hobbs R, Suding K (eds) New models for ecosystem dynamics and restoration, Washington DC: Island Press. pp 3-21
34. Thomas R (2009) Regent Honeyeater habitat restoration project Lurg Hills, Victoria. Ecol Manage Restor 10(2):84-97
35. Thomas R (2012) Overall achievements. Regent Honeyeater Newsl Summer. pp 2011-12
36. Walker B, Salt D (2006) Living in a complex world: an introduction to resilience thinking. In: Resilience thinking: sustaining ecosystems and people in a changing world. Washington DC: Island Press. pp 1-14
37. Wilkins S, Keith DA, Adam P (2003) Measuring success: evaluating the restoration of a grassy eucalypt woodland on the Cumberland Plain, Sydney, Australia. Restor Ecol 11(4):489-503

38. Young TP, Peterson DA, Clary JJ (2005) The ecology of restoration: historical links, emerging issues and unexplored realms. Ecol Lett 8:662-673 Publisher Full Text
39. Zedler JB, Doherty JM, Miller NA (2012) Shifting restoration policy to address landscape change, novel ecosystems, and monitoring. Ecol Soc 17(4): 10.5751/ES-05197-170436

Chapter 7

Life Cycle Assessment of Producer Gas Derived from Coconut Shell and its Comparison with Coal Gas: An Indian Perspective

Chenicheri Chandroth Sreejith, Chandrasekharan Muraleedharan, and Palatel Arun

Department of Mechanical Engineering, National Institute of Technology Calicut, NIT Campus (P.O.), Kozhikode, 673601, India

ABSTRACT

Oil dependency and global warming have stimulated R&D activities on the utilization of secondary biofuels. This article investigates the suitability of coconut shell-derived producer gas (a secondary gaseous biofuel) as a substitute for coal gas from an environmental perspective using life cycle assessment (LCA) approach. Thermochemical gasification in an air-fluidized bed with steam injection is the gaseous fuel production process. LCA is carried out with respect to Indian conditions based on primary and modified Ecoinvent 2.0 data using IMPACT 2002+ environmental impact assessment methodology. The study indicates that coconut shell-derived producer gas life cycle is capable of saving 18.3% of emissions causing global warming potential, 64.1% of emissions causing ozone depletion potential, and 71.5% of nonrenewable energy consumption. The analysis of energy and exergy consumptions of the two life cycles reveal that the renewable fraction in the total energy demand is 62.9% for producer gas life cycle, while it is only 2.8% for coal gas life cycle. Allocation of the by-product, coconut palm residues, is the major responsible factor for this reduction in environmental burden. Based on the existing fertilization practice and the utilization of electricity from the Indian electric grid, substantial green house gas (GHG) emission savings cannot be achieved for producer gas life cycle over coal gas life cycle. However, the green electricity-aided catalytic gasification of coconut shell produced by organic farming can result in 43.4% reduction in GHG emissions, meeting European Union renewable energy directive.

BACKGROUND

Life cycle assessment (LCA) is a process of evaluating environmental burdens or benefits associated with the total life cycle of a product. This is conducted by identifying and quantifying the energy and materials used and waste products released into the environment [1]. It is performed in accordance with ISO 14040:2006 and ISO 14044:2006 standards. In simple terms, LCA is a holistic tool for evaluating the environmental impacts associated with a product, process, or activity. The ecological or environmental impacts include effects on the ecosystem, human health, and natural resources [2]. Generally, LCA is termed as a 'cradle to grave' approach since it accounts the environmental impacts throughout the product's life, right from raw material acquisition, processing, manufacturing, use, and finally its disposal [3]. The significance of LCA lies in the fact that it equips the policy makers and decision makers for adoption of suitable and sustainable energy supply systems. Increasing global concern due to air pollution and to limited oil reserves has generated much interest in environmental friendly alternatives to petroleum-based fuels [4]. This is more prominent in the context of the current substantial contribution of fossil fuels (over 85%) [5] to the global energy supply. Generally, biofuels produced from renewable energy sources (biomass) are considered to be an effective alternative to fossil fuels, and the concerned production technology is sustainable. To identify savings in energy and emissions from biofuel production and utilization, a thorough evaluation of the corresponding life cycle is to be carried out carefully [6]. LCA is an effective tool for this, which accounts for the relative environmental impacts of biofuel life cycle with respect to 'base case' such as fossil fuel-based life cycle. For example, for biodiesel, the base case may be petroleum diesel, and for producer gas derived from biomass, the base case may be coal gas. The adoption of LCA in the perspective of sustainable economy is more essential nowadays because of the boom in the installation of energy and heat production systems utilizing these sustainable energy techniques with the intention of meeting European Union (EU) targets of 20% renewable energy share by 2020.

The updated standard ISO 14044:2006 [7] specifies requirements and provides guidelines for life cycle assessment including definition of goal and scope of the LCA, life cycle inventory analysis (LCI) phase, life cycle impact assessment phase, life cycle interpretation phase, reporting critical review of the LCA, limitations of the LCA, relationship between the LCA phases, and conditions for use of value choices and optional elements.

LCA approach was adopted for assessing the environmental suitability of the utilization of various renewable energy resources such as solar PV, wind, and biomass by Manish et al.[8]. The base cases considered were coal-based thermal power generation and steam methane reforming for hydrogen production. The comparison of different options was made based on major indicators like life cycle cost, green house gas (GHG) emissions, and net energy ratio (NER). Secondary biofuels produced from unprocessed primary biomass-based materials are sustainable alternative to fossil fuels. Secondary biofuels may be of solid, liquid, or gaseous types which are produced by adopting various production routes. A state-of-the-art review of literature pertaining to liquid biofuels was reported by Nigam and Singh [9]. Producer gas is a secondary gaseous biofuel generated by the thermochemical gasification of biomass. NER and total energy ratio of the life cycle of rice straw-generated producer gas were evaluated by Shie et al. [10]. A comparison of biomass and coal in terms of several environmental indicators for integrated gasification combined cycles with [11] and without [12,13] CO_2 chemical absorption revealed the superiority of biomass. On further analysis of the potential of biomass to produce hydrogen, Koroneos et al.[2] reported the advantage of biomass gasification-syngas reformation-absorption route over biomass gasification-electricity generation-electrolysis route. Manish et al. [8] on reviewing the LCA studies on gasification with biomass and with biomass-coal blends for electricity generation reported an evident reduction in environmental impacts when biomass is used as energy source. However, the life cycle of biomass-based systems can have higher eutrophication potential [14] and acidifying emission potential. Consequences of using agricultural land for other purposes than food production [15] and the size and scale of the biomass-based combined heat and power (CHP) plant [16] are significant factors while utilizing biomass-based systems.

The choice of selected indicators for LCA can provide options for the improvement of the existing systems. Such a damage assessment based on the existing life cycle of natural gas combustion district heating system at a rural location in British Columbia was conducted by Pa et al.[17,18] recently. Similar region-specific LCA studies for woodchip-based 'green electricity generation' in Austria [1], post-consumer wood and forest residue-based gasification in a metropolitan area in Barcelona [19], and carbonization of several woody biomass (found plenty in Singapore) to charcoal [20] are worth mentioning. The comparatively lesser impacts on global warming, acidification, and eutrophication by the life cycle of biomass thermal technologies over that of biological routes were reported by Zaman [21] and Zhong et al. [22]. Iribarren et al. [23], on considering biofuel production systems from poplar biomass, stressed on the need for plantation and harvesting of the lignocellulosic biomass in terms of land occupation and fertilizer requirements. The utilization of hydrogen as fuels in vehicles and the progress in fuel cells have necessitated the need for LCA of the concerned hydrogen supply chains. Such analyses were reported for urban transportation with hydrogen vehicles [24] and for hydrogen refueling stations[25]. A comprehensive state-of-the-art review of environmental impacts of hybrid and electric vehicles was reported by Hawkins et al. [26].

Present Study and its Significance

India, one of the biggest agricultural countries in world, is in the process of setting up many large-and medium-scale biomass-based power plants for rural electrification, utilizing locally available feedstocks. Thermochemical gasification is a mature sustainable technology for gaseous fuel production from renewables [27,28]. The country has declared its commitments to the implementation of clean development mechanism, which is an arrangement under the Kyoto protocol. The abundantly available biomass materials in the subcontinent include eucalyptus wood, rice husk, paddy husk, coconut shell, cashew nut shell, and bamboo. The climatic conditions are not uniform in the country, making it suitable for these agricultural and forest biomass to grow in different parts of the country. For example, in the southern part of India, the most abundantly available biomass feedstock is coconut shell. Thus, utilization of coconut shells for bioenergy production for electrifying rural areas in southern India has significance. The environmental burdens associated with the utilization of these biomasses (and more) are largely different since irrigation schedule (depends on rain fall), fertilization schedule (depends on soil conditions), land requirement, yield of biomass per hectare, and heating value are different for these biomasses. Thus, it is vital to make assessments about various biomass feedstocks suitable for such power plants from an energy and environmental perspective. As a first step in the venture, coconut shell is taken for assessment in the present study, which will be followed by several other probable feedstocks mentioned already. Hence, the present work is a region-specific study for LCA of coconut shell-derived producer gas life cycle in the third largest coconut-producing country in the world. To the best of the authors' knowledge, no work has been reported on a comparative LCA of coconut shell-derived producer gas and coal-derived coal gas with respect to Indian conditions.

As the world is heading towards hydrogen energy economy, many attempts were reported for enhancing the hydrogen concentration in biomass-derived producer gas for the onward utilization in fuel cells and in hydrogen-fuelled vehicles. Steam reforming the producer gas generated by air-fluidized bed (AFB) gasification is one of the options for achieving this. Syngas (producer gas) can be made further rich in hydrogen by supplying more amount of steam to the AFB gasifier.

However, the most common production method of steam utilizes fossil fuels which generate greenhouse gases and ozone-depleting gases, necessitating the optimization of steam utilization. Thus, the present study is significant as a 'well-to-tank' LCA approach to make comparison between two solid fuel-hydrogen-rich gaseous fuel life cycles, that is, the life cycles of fossil fuel-derived coal gas and its potential renewable alternative (coconut shell-derived syngas) are compared with respect to India, the world's largest low-ash-content coconut shell producer.

Location of the Present Study

Kerala, one of the southern states of India, is blessed with abundant rain fall in the range of 2,000 to 3,000 mm/year [29]. The fertile soil conditions in the region are well suited for coconut cultivation. Researches on a variety of coconut palm species have succeeded in developing new hybrids suiting varying conditions [30]. The specialty of coconut palm is that every part of it has valuable applications, as given in Table 1. The major use of coconut shell (main product) is for handicrafts and charcoal, the latter being one of the base materials for the manufacture of activated carbon. Coconut shell is also used as a primary biofuel for open combustion. The co-product (similar revenue to the main product) coconut oil can be used for edible and inedible (lubricant with nanoparticles addition) purposes. The by-product (lower revenue than that of the main product) palm residues find utilization as a fuel with energy content of 16 to 18 MJ/kg. The state is in the process of developing medium-scale gasification plants for rural electrification utilizing coconut shells as feedstocks for the past 2 to 3 years. The present study pertains to the conversion of coconut shell (primary biofuel) to producer gas (secondary biofuel), which has not yet been analyzed in this region on a life cycle basis.

Table 1: Products, co-products, and by-products during coconut shell production

Category	Item	Uses
Product	Coconut shell	Charcoal, handicrafts, fuel for open combustion
Co-product	Coconut oil	Edible oil, lubricant

By-product (coconut palm residues)	Petioles and leaf	Fuel, hut thatching, organic manure
	Coconut seed cake	Cattle feed, organic manure
	Coconut husk	Fuel, organic manure, raw material for coir
	Stem	Piles, pillars, fuel

Sreejith *et al.*

Sreejith *et al. International Journal of Energy and Environmental Engineering* 2013 4:8, doi:10.1186/2251-6832-4-8

METHODS

Life Cycle Inventory Analysis

Goal, Scope, and Functional Unit

The major objective of the present analysis is to compare the environmental impacts of coconut shell-derived producer gas life cycle with that of the coal gas (generated by the gasification of coal) life cycle. Thus, the present LCA aims for a comparison of the relative environmental impacts of renewable and nonrenewable alternatives of energy production and quantifies the benefits from replacement of the latter with the former. In this study, contributions of various LCA phases to the overall environmental impacts are analyzed to identify the most significant phase in the life cycle. The reference functional unit for the study is 1 MJ energy content in the gaseous fuel (producer gas generated from coconut shell gasification or coal gas generated from coal gasification). This functional unit is selected since the energy content may be utilized for various purposes such as fueling transport system, production of electricity, and as a source of heat for systems like absorption refrigeration.

Producer Gas Life Cycle

System Boundaries

The coconut shell-derived producer gas life cycle is analyzed in the perspective of the agricultural scenario existing in the state of Kerala in India. The geographic boundary for the fossil fuel reference system is India except where resources are extracted and transported to India from other countries. A well-to-tank approach is adopted for the LCA because of the potential of the energy content of the generated gas for various purposes, as mentioned in the section 'Goal, scope, and functional unit'.

The life cycle of the producer gas is presented schematically in Figure 1, including crushed coconut shell production and gasification. The various important subsystems include sand bed preparation, coconut cultivation, fertilizing, irrigation, coconut shell production, shell crushing, generation of gasification agents, and coconut shell gasification in a FB gasifier.

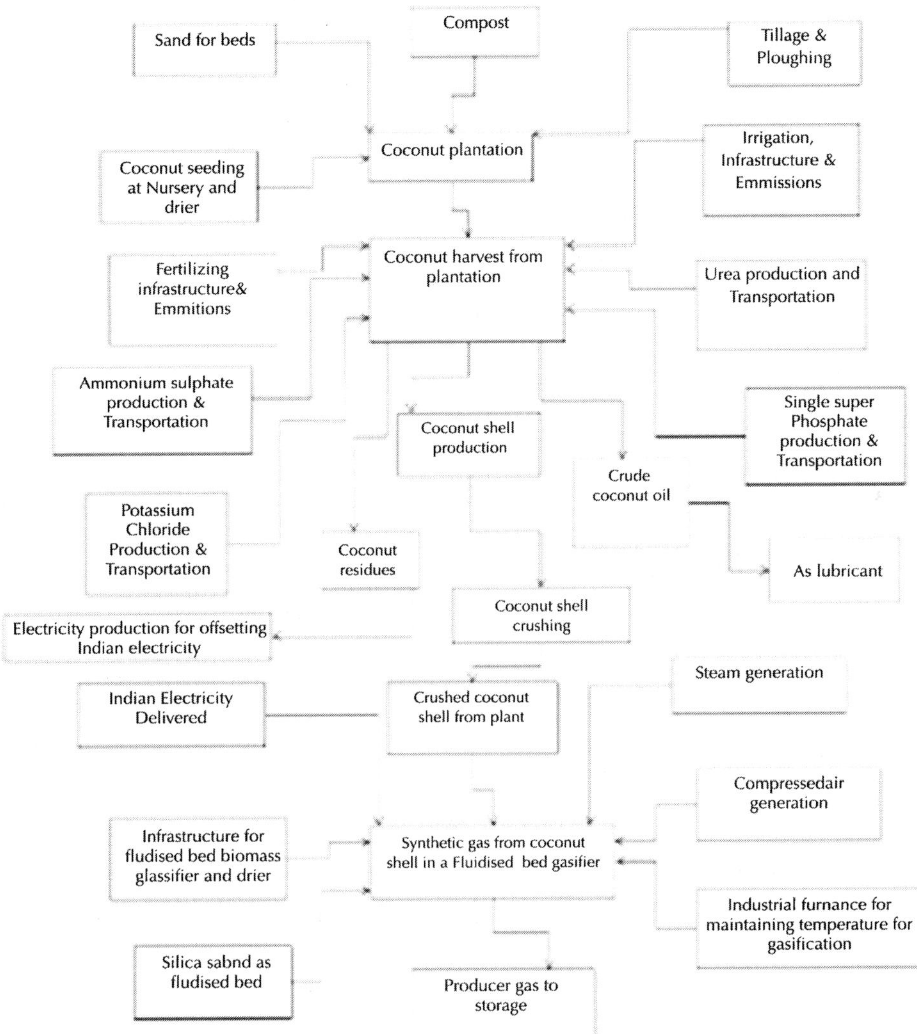

Figure 1: Schematic of producer gas life cycle.

Coconut Cultivation

Coconut palms are generally cultivated in rain-fed conditions and in irrigated conditions. Irrigation is mainly required in the summer season at predetermined intervals. Average rain fall in Kerala based on the data [29] available for the past 50 years is shown in Table 2. Based on the data, the duration of the summer season is taken as 3.5 months, and irrigation is required in this season. The common schedule [31] of irrigation is represented in Table 3.

Table 2: Average monthly rain fall in Kerala

Month	Rainfall (mm)
January	14.6
February	16.6
March	36.1
April	110.9
May	252.6
June	653.2
July	687.2
August	404.7
September	252.3
October	270.7
November	158.6
December	45.9

From [27].

Sreejith *et al.*

Sreejith *et al. International Journal of Energy and Environmental Engineering* 2013 4:8, doi:10.1186/2251-6832-4-8

Table 3: Irrigation schedule for coconut cultivation

Age of palm	Quantity of irrigation (Liters per palm)	Schedule (in the summer season, 3.5 months)
First 2 years	1,110	Once in 4 days
Third year onwards	600 to 800	Once in 7 days

From [29].

Sreejith *et al.*

Sreejith *et al. International Journal of Energy and Environmental Engineering* 2013 4:8, doi:10.1186/2251-6832-4-8

Inventory Data for Coconut Cultivation

The following data [31] for coconut cultivation are applied for the development of the sub-modules: sand bed preparation, coconut seedling, coconut plantation, irrigation, and fertilizing. Crude coconut oil and palm residues are the co-product and by-product, respectively, in the coconut shell production process.

- Coconut seedlings are grown in a nursery as per established specifications of planting.
- Seedling survival rate is assumed as 65%, and the percentage of healthy seedlings which can be replanted is taken as 75% as per data from local farmers.
- Selected seedlings are planted in pits of 1-m^3 size, and 5 kg of organic manure is used.
- The tree density is 175 trees/ha.
- Irrigation and fertilization schedules are according to the principle of good management practices [31], as given in Tables 3 and 4, respectively.
- The coconut yield from a palm per year is 125 nuts, with an average weight of 1.6 kg for full nut and 0.5 kg for husked nut.
- The average quantity of dried copra from nuts is in the rate of 18 kg from 100 nuts (or from 50 kg of husked nut).
- The oil content of copra is about 70%, and the extraction efficiency is about 85%. Hence, expected oil recovery is about 60% of the weight of copra.
- • Based on the above conditions, 4.7 kg of husked nuts (9.4 numbers equals approximately 10) is required to produce 1 kg of coconut oil.
- Also, 0.64 kg of shell is obtained from 1 kg of husked nut. It means that from a palm per year, 40 kg of nut shell can be obtained. Thus, 3.1 numbers of husked nuts are required to produce 1 kg of shell.
- Palm residues include the leaves and the dried leaf weighing 2 kg per tree on average and has a heating value of 16,500 kJ/kg.

Table 4: Fertilization schedule for coconut plantations according to the principles of good management practices

Age of coconut palm	Nutrient dosage	Quantity of fertilizer to be added (g)			
		Ammonium sulfate	Urea	Single super phosphate	Muriate of potash
3 months	1/10 of full dose	250	110	180	200
First year	1/3 of full dose	800	360	590	670
Second year	2/3 of full dose	1,675	720	1,180	1,340
Third year onwards	Full dose	2,500	1,080	1,780	2,010

From [29].

Sreejith *et al.*

Sreejith *et al. International Journal of Energy and Environmental Engineering* 2013 4:8, doi:10.1186/2251-6832-4-8

In the coconut cultivation phase, chemical processes for the production of fertilizers, the utilization of these substances, physical infrastructure for coconut cultivation, and emissions from fertilizers and irrigation are included in the system boundary. Thus, almost all the sub-modules under the modules such as seedling, irrigation, composting, and fertilization in coconut cultivation are based on primary data. For fertilizer production, Ecoinvent 2.0 data were modified with respect to Indian electricity generation and transmission.

Coconut Shell Gasification

The technology for gaseous fuel production from solid feedstocks (coal or coconut shell) considered in this study is thermochemical gasification. Producer gas or coal gas with more hydrogen concentration is generated by steam injection into an AFB gasifier as mentioned in the section 'Present study and its significance'.

A gasification module is developed for the thermochemical treatment of crushed coconut shell. The producer gas production phase consists of sub-modules such as steam and compressed-air generation, sand preparation for its utilization as bed material, physical infrastructure for the gasification plant, and electric furnace for maintaining the reactor temperature. A carbon conversion efficiency of 100% for gasification is assumed, and the by-products of gasification (tar and ash) are neglected. The energy requirements (data modified with respect to Indian electricity mix) for all the relevant processes such as fertilizer production, coconut crushing, steam and compressed-air production, irrigation, and thermochemical gasification are included while developing the model.

Coal Gas Life Cycle

Coal gas life cycle is used as the reference system for comparison with that of coconut shell-derived producer gas life cycle. The important sub-modules involved are coal transport from foreign mines, coal mining in India from open and underground mines, transport of Indian and imported coal to regional storage, and gasification of this coal mix to generate coal gas. India is the third biggest hard coal producer (526 million tonne/year) after China (2,971 million tonne/year) and the USA (919 million tonne/year). Eighty-five percent of the total coal utilization in the country is from domestic production, and the remaining 15% is imported from countries like Australia, South Africa, and North America [32]. Out of the domestic production, 85% is from open mining, and the remaining is from underground mining. Major domestic coal mining companies are Coal India Limited, Neyveli Lignite Corporation, and Singareni Collieries Limited. Thus, domestic coal is considered to be mined by these companies at locations in Dhanbad, Neyveli, and Hyderabad, respectively. The study pertains to Calicut, a northern city in Kerala, which is the regional storage location taken for modeling of the transport module for coal (imported and domestic). The mode of transport of coal to regional storage location and the corresponding distances are given in Table 5. For coal transport, the Ecoinvent database is modified with respect to Indian electricity consumption for diesel production for fuelling the transport system (rail, road). The transports of the imported coal in other countries are modeled using the respective data source available in the Ecoinvent 2.0 database. Figure 2 shows the coal gas life cycle map.

Table 5: Mode of coal transport and distance to central location

	Mode of transport	Distance (km)	Remarks
Imported coal	Sea	36,336	Total distance from the three countries to the central storage location
	Rail	1,212	
	Road	110	
Domestic coal	Rail	3,994	Total distance from three domestic mining locations in India to the central storage location
	Road	377	

Sreejith *et al.*

Sreejith *et al. International Journal of Energy and Environmental Engineering* 2013 4:8, doi:10.1186/2251-6832-4-8

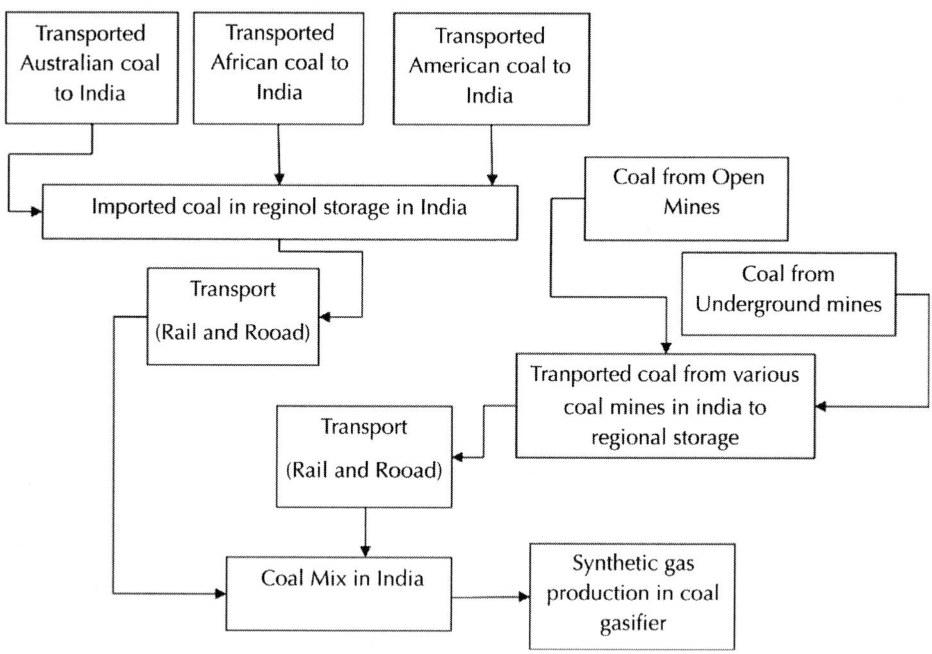

Figure 2: Schematic diagram of coal gas life cycle.

Supporting Process

A schematic of Indian electricity generation is shown in Figure 3. India's specific electricity generation module was created based on the national average annual proportion of electricity generation by fuel type. LCA modules were developed with contributions of 12%, 70%, 2%, 1%, and 15% of the gross electricity generation [33] in the country by natural gas, coal, nuclear, renewable, and hydropower plants, respectively. Due to the unavailability of Indian data in the Ecoinvent 2.0 database, western European data were modified with respect to Indian electricity share for various energy sources.

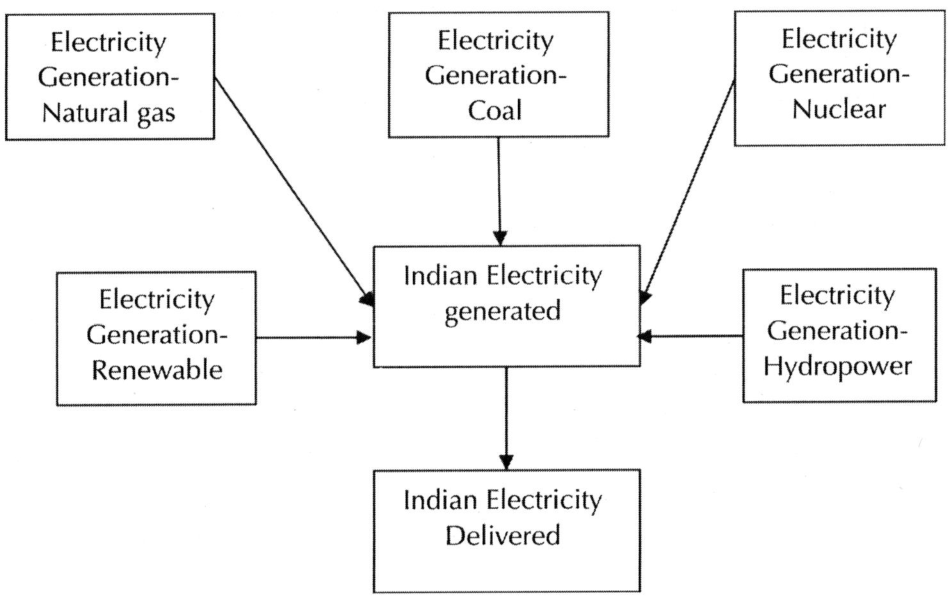

Figure 3: Modeling schematic for Indian electricity generation.

Similarly, the fuel data (coal and natural gas) for the power plants in the Ecoinvent database were modified incorporating the proportion of domestic production and import. The transport data were also modified accordingly.

Allocation Procedure

Allocation is necessary when a process produces more than one valuable product. In such cases, it is necessary to divide the environmental impacts from the process between the products. Crude coconut oil and coconut residues are the co-product and by-product, respectively, of coconut shell production process. The utilization of energy in the coconut palm residues (by-product of coconut cultivation stage) and production of coconut oil are two major cases in the present LCA study. The 'division' of the environmental impacts may be done either by allocation or by system expansion. According to Gnansounou et al. [34], environmental impacts in a life cycle may even vary with the allocation method (mass, energy, carbon content, or economy) adopted. In this study, system boundary expansion method is adopted for allocating the co-product and by-product.

Agricultural residues of coconut plantation play a significant role in the energy analysis because of its availability in large quantity and its energy content. As mentioned in 'Inventory data for coconut cultivation', palm residues include the leaves and the dried leaf weighing 2 kg/tree on average and have a heating value of 16,500 kJ/kg. The tree density is 175 palm/ha. The thermal energy that can be generated from the palm residues is calculated from its lower heating value (LHV), followed by the computation of its conversion into electrical energy with an overall conversion efficiency (from palm residues to electric energy) of 25%. This assumption does not take into account the technology used for this conversion. Based on the inventory data for coconut cultivation provided in 'Inventory data for coconut cultivation', the net electricity generated from coconut palm residues is computed for the reference functional unit (1 MJ) in SimaPro. This amount of electricity is assumed to offset the corresponding Indian electricity production from coal and natural gas. Thus, the system boundary (as explained in 'System boundaries') is expanded to include the system of processes which are involved in the production of Indian electricity from coal and natural gas.

The environmental burdens are allocated for coconut oil by assuming it replacing commercial paraffin oil lubricant [35]. As mentioned in 'Inventory data for coconut cultivation', 4.7 kg of husked nuts (9.4 numbers equals approximately 10) is required to produce 1 kg (or 1.08 l) of coconut oil. Thus, corresponding to the generation of 1 MJ of syngas energy (reference functional unit), the net quantity of 'lubricant' (coconut oil) production can be computed and modeled in SimaPro. A separate module is created for paraffin oil lubricant production (modified Ecoinvent module), and the producer gas life cycle system boundary is expanded to include the system of processes in the commercial paraffin oil lubricant production. By proceeding like this, the net emissions associated with the production of producer gas with 1 MJ of energy are determined by subtracting the sum of the emissions (or impacts) associated with the saved fossil fuel-based electricity production and the saved commercial lubricant production from the emissions associated with the generation of producer gas.

Life Cycle Impact Assessment

Following cutoff criteria are adopted in the study:

a. Since the maturity duration for coconut palm is 10 years, the study is conducted to analyze the impacts of producer gas life cycle corresponding to this duration.
b. Application of pesticides, insecticides, and herbicides for coconut cultivation is avoided due to lack of readily available local data.
c. CO_2 emissions from the utilization of producer gas are taken as nullified by the CO_2 absorption during the growth of coconut tree. This cutoff criterion is made in the context of unavailability of a well-established database to define the rate of sequestration of carbon by coconut palms.
d. Potential land use changes are not evaluated assuming the plantations are long-term existing coconut plantations.
e. Transport of personnel to carry out the different stages in the life cycle is not included.

Environmental impact assessment of the two life cycles is conducted by computational implementation of the inventories collected, using the software SimaPro 7.3.3® developed by Product Ecology Consultants (Amersfoort, The Netherlands) [36]. This software is widely used[8,12,21] for the inventory analysis and LCA of chemical and energy system chains. All the modules of the life cycle are developed using the Ecoinvent 2.0 database available in SimaPro. Tables 6 and 7 represent the corresponding module descriptions in SimaPro for producer gas life cycle and coal gas life cycle, respectively. Table 8 provides details of the inventory data (primary and sourced from Ecoinvent) collected for model development. The impact assessment method selected is IMPACT 2002+ (IMPact Assessment of Chemical Toxics) available in the SimaPro methods library. IMAPCT 2002+ methodology links the LCI results to four broad damage categories: human health, ecosystem quality, climate change, and resources. The schematic of the IMPACT 2002+ framework [36] is shown in Figure 4. The following three specific environmental impact categories are selected in this study. Brief descriptions on the mentioned impact categories are given in the Appendix.

1. Global warming potential (GWP): Net emission of Intergovernmental Panel on Climate Change (IPCC)-identified greenhouse gases are calculated and expressed in terms of kilograms of CO_2equivalent for a 100-year lifetime.
2. Ozone depletion potential (ODP): The emission of gases, mainly CFC-10, halon 1301, and halon 1211, is calculated and expressed in kilograms of CFC 11 equivalent.
3. Fossil fuel depletion potential (FDP): The degree of fossil fuel consumption is determined and expressed in terms of megajoules of energy surplus.

Table 6: SimaPro module descriptions for coconut cultivation and shell gasification in a FB gasifier

Module	Purpose
Sand for beds	Ecoinvent 2.0 module for production and related emissions of sand to be used in the nursery for coconut seedlings
Compost	Custom module for production and related emissions of compost to be used in planting coconut seedlings
Tillage and plowing	Modified Ecoinvent 2.0 module for land preparation for plantation and related emissions. Diesel source modified to Indian conditions
Coconut seedling at nursery	Custom module to represent the requirements for coconut seedlings at nursery

Coconut plantation	Custom module built from three inputs. Tillage, seedlings, and compost
Fertilizing: infrastructure and emissions	Modified Ecoinvent 2.0 module representing impacts of fertilizing 1 ha of plantation using the required machinery
Irrigation, infrastructure, and emissions	Modified Ecoinvent 2.0 module representing impacts of irrigating 1 ha of plantation using the required machinery
Ammonium sulfate production and transportation	Ecoinvent 2.0 module representing impacts of fertilizer production for usage in 1 ha of plantation over the lifetime
Urea production and transportation	Ecoinvent 2.0 module representing impacts of fertilizer production for usage in 1 ha of plantation over the lifetime
Potassium chloride production and transportation	Ecoinvent 2.0 module representing impacts of fertilizer production for usage in 1 ha of plantation over the lifetime
Single super phosphate production and transportation	Ecoinvent 2.0 module representing impacts of fertilizer production for usage in 1 ha of plantation over the lifetime
Coconut harvested from plantation	Custom module representing impacts associated with managing and operating the plantation. Energy from the combustion of biomass residues is used to offset Indian electricity delivered.
Coconut shell production	A custom module to represent the impacts associated with coconut shell production from the harvested coconuts
Crude coconut oil	Impacts of crude coconut oil production and related emissions in an oil extraction plant. Modified Ecoinvent 2.0 module is used with coconut harvesting and electricity utilization corrected to suit Indian conditions.
	Crude coconut oil (by-product) is allocated as lubricant for machinery with the addition of CuO nanoparticles.

Crushing of the shell	Impacts of grinding the coconut shell in biomass crusher for utilization in FB gasifier. Modified Ecoinvent 2.0 module is used with electricity utilization and input material quantity modified to suit Indian conditions.
Coconut shell gasification	Modified Ecoinvent 2.0 module for assessing the impact associated with crushed coconut shell gasification in a FB gasifier with steam and air as gasification medium
Steam for gasification	Modified Ecoinvent 2.0 module for steam generation. Modified with respect to Indian electricity mix
Compressed air for gasification	Modified Ecoinvent module for the assessment of impacts of producing compressed air at 2 bar pressure for fluidizing the bed

Sreejith *et al.*

Sreejith *et al. International Journal of Energy and Environmental Engineering* 2013 4:8, doi:10.1186/2251-6832-4-8

Table 7: SimaPro module descriptions for coal mining and gasification

Module	Purpose
Silica sand in the gasifier plant	Ecoinvent 2.0 module for production and related emissions of sand (SiO2) to be used as bed material including the energy for drying the sand
Indian electricity delivered	Custom module to represent the impact of production and delivery of electricity for coal gasifier plant operation. The proportion of electricity generation is as given in Figure 4.
Steam generation at the site	Modified Ecoinvent 2.0 module for steam generation and related emissions. Electricity production and delivery modified to Indian conditions

Compressed-air generation	Modified Ecoinvent 2.0 module for compressed-air generation supply. The module includes a screw-type compressor, lubricating oil, electricity delivery, and transport of compressor and lubricating oil to the installation site.
Infrastructure for coal gasifier	Ecoinvent 2.0 module representing impacts of developing infrastructure for a typical coal gasifier. The dataset includes land use, building and facilities, dryer, gasifier, and gas treatment facility.
Industrial furnace for maintaining the temperature of the gasifier	Ecoinvent 2.0 module representing impacts of an oil furnace of 1-MW capacity. The module includes the production of oil, the infrastructure for the oil boiler, and transport to site.
Domestic coal mining and transport	Modified Ecoinvent 2.0 module representing impacts of domestic mining (underground and open) in India and its transport to a regional storage
Imported coal	Modified Ecoinvent 2.0 module representing impacts of coal imports from three countries, South America, South Africa, and Australia, in the ratio 2:1:7 to the regional storage location in India. Built-in Ecoinvent module for coal mining and transport in the respective countries were taken into account.
Coal transport to the plant site	Modified Ecoinvent 2.0 module including the impacts of road and rail transport in India.
Synthetic gas production in coal gasifier	Modified Ecoinvent 2.0 module representing impacts of gas production in fluidized bed coal gasifier with steam and air as gasifying medium

Sreejith *et al.*

Sreejith *et al. International Journal of Energy and Environmental Engineering* 2013 4:8, doi:10.1186/2251-6832-4-8

Table 8: Details of inventory data

Process/module	Data type and representativity to local conditions
Electricity generation	Primary data
	Electricity production mix is as follows: coal 70%, natural gas 12%, hydropower 15%, nuclear 2%,renewable 1%
	For each plant type, the import and domestic production of the corresponding fuel in India is accounted.
	For example, for coal power plants, the import/ domestic production ratio in India is 8.5:1.5, whereas for the Ecoinvent European database, the ratio is 0.03:9.97.
	Similar modification is applied to other energy sources for a better representation of Indian conditions.
	For hydropower plants, the production capacity (kWh) is modified as the ratio between run-off river plant and reservoir plant in the ratio 0.2:0.8
Transport of coal	Modified Ecoinvent data
	For transport of fuels to a central location, the modules in the Ecoinvent database are modified incorporating the average distances by rail, road, and sea, as given in Table 5. Ecoinvent data are utilized for the development of modules for coal transport in the three countries from where coal is imported.
Infrastructure for electricity generation and transmission loss	Modified Ecoinvent data
	For the sub-process, infrastructure for electricity generation, the Ecoinvent database is modified by changing the land area required for each type of power plant for the process.
	For the sub-process transmission loss, modification is made in terms of Indian transmission loss of 35%.

Steam and compressed-air generation for gasification	Modified Ecoinvent data
	Western European data in the Ecoinvent database are modified with respect to electricity production and delivery in India.
Infrastructure for biomass gasification plant	Modified Ecoinvent data
	The Swiss database is modified with Indian electricity supply and vehicle transport (rail, lorry) suitable to Indian conditions.
Coconut shell production and crushing	Primary data
	A primary module with all the sub-modules developed based on the data corresponding to southern Indian conditions. The sub-modules include coconut plantation, irrigation, fertilization, composting, sand mix, tillage and plowing, crushing, etc. Allocation is made for crude coconut oil and coconut palm residues.

Sreejith *et al.*

Sreejith *et al. International Journal of Energy and Environmental Engineering* 2013 4:8, doi:10.1186/2251-6832-4-8

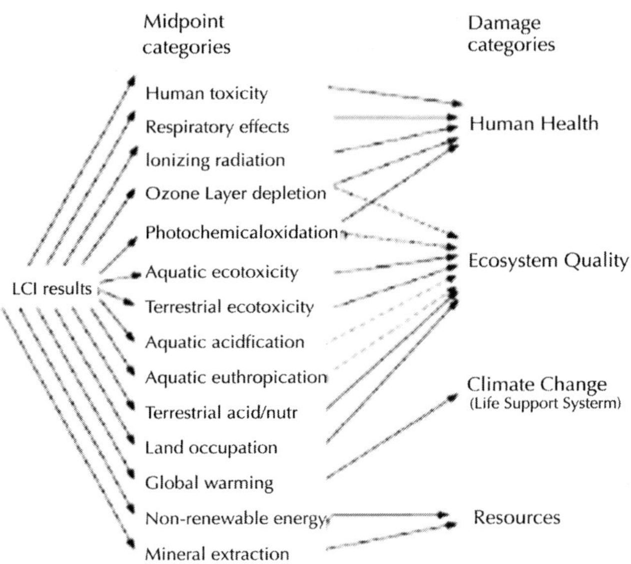

Figure 4: Overall scheme of IMPACT 2002+ framework.

RESULTS AND DISCUSSION

In the following paragraphs, impact assessment results of the two life cycles are discussed in terms of the corresponding normalized unit for each impact category. The comparison between the life cycles and contributions of various sub-processes to these impacts are also discussed.

Global Warming Potential

A 100-year GWP value in kilograms of CO_2 equivalent, normalized by the functional unit of the study, i.e.,1 MJ heating value, is presented in Table 9. Results show that a saving of 18.3% in GHG emissions causing GWP can be obtained for the producer gas life cycle over coal gas life cycle. Process contributions of GHG emissions causing GWP in the two life cycles are given in Tables 10and 11. From these tables, it is evident for the producer gas life cycle that the major contributions are from (1) steam generation, (2) coconut oil production, (3) fertilization, and (4) irrigation. Utilization of fossil fuel-based Indian electricity for steam generation and irrigation purposes is responsible for the emissions of greenhouse gases such as CO_2. The major source of potential emissions of fertilizer application is nitrogen oxides, computed according to Equation 1 [37].

$$N_2O_{direct} = \left[\sum_i {}_{mF,i.} NC_{F,i}(1-Frac_{gasf})\right]$$
$$EF.mw_{N_2O.GW_{N_2O}} \tag{1}$$

where N_2O_{direct} is the direct N_2O emissions (tonne CO_2 equivalent), $m_{F,}i$ is the mass of synthetic type i fertilizer applied (tonne), $NC_{F,}i$ is the nitrogen content of synthetic type i fertilizer applied (gN/100 g fertilizer), $Frac_{gasf}$ is the fraction that volatilizes as NH_3 and NO*X* for fertilizers (dimensionless), EF is the emission factor for emissions (tonne N_2O), mwN_2O is the ratio of the molecular weights of N_2O and N (tonne N_2O/tonne N), and GWN_2O is the global warming potential for N_2O (kg CO_2 equivalent/kg N_2O).

Table 9: Impact assessment of the two life cycles

Impact	Unit	Coal gas life cycle	Coconut shell-derived producer gas life cycle	Percentage of decrease for producer gas life cycle
GWP due to GHG emissions	kg CO2 equivalent/ MJ energy	0.133	0.108	18.3
Emissions causing ODP	kg CFC 11 equivalent/ MJ energy	2.39E−8	8.6E−9	64.1
FDP	MJ energy surplus/MJ energy	4.19	1.4	71.5

Sreejith *et al.*

Sreejith *et al. International Journal of Energy and Environmental Engineering* 2013 4:8, doi:10.1186/2251-6832-4-8

Table 10: Process contributions for producer gas life cycle

Contributing life cycle process	GWP (kg CO2equivalent/ MJ energy)	ODP (kg CFC 11 equivalent/ MJ energy)	FDP (MJ energy surplus/MJ energy)
Gasification of coconut shell	1.21E−6	0	0
Electricity production and supply for gasification	6.2E−3	4.82E−10	0.0806
Steam production and supply for gasification	0.2068	8.3E−9	1.30
Compressed-air generation and supply for gasification	8.56E−3	6.64E−10	0.1116

Gasifier infrastructure establishment	8.06E–5	7.4E–12	1.15E–3
Crushing coconut in machine	4.16E–6	2.0E–12	6.58E–5
Electricity for crushing mill operation	2.45E–4	1.9E–11	3.18E–3
Crude coconut oil production (co-product allocation)	–0.098	–1.35E–9	–0.24
Fertilization machinery	2.53E–3	3.68E–10	0.04105
Irrigation	0.098	5.0E–9	0.98
Fertilizers	0.1009	1.19E–8	1.75
Coconut plantation (seedling, tillage, and compost)	2.89E–4	3.29E–11	3.6E–3
Electrical energy to main grid (by-product allocation)	–0.324	–9.46E09	–2.89
Furnace for gasifier operation	3.8E–6	2.46E–13	1.07E–4
Bed material, catalysts, etc.	1.86E–3	1.88E–10	0.034
Transport and maintenance of all the vehicle transport infrastructure	1.57E–3	2.6E–10	0.02701
Disposal of ash and sewage treatment	4.24E–5	1.006E–12	1.37E–6

Sreejith *et al.*

Sreejith *et al. International Journal of Energy and Environmental Engineering* 2013 4:8, doi:10.1186/2251-6832-4-8

Table 11: Process contributions for coal gas life cycle

Contributing life cycle process	GWP (kg CO2equivalent/MJ energy)	ODP (kg CFC 11 equivalent/ MJ energy)	FDP (MJ energy surplus/ MJ energy)
Gasification of coal	1.09E−6	0	0
Electricity production and supply for gasification	5.58E−3	4.338E−10	0.07254
Steam production and supply for gasification	0.09612	7.47E−09	1.251
Compressed-air generation and supply for gasification	7.70E−3	5.976E−10	0.1004
Gasifier infrastructure establishment	7.25E−5	6.66E−12	1.04E−3
Imported coal			
Road transport	9.54E−5	1.53E−10	1.701E−3
Rail transport	1.13E−3	9.18E−10	0.021474
Electricity	5.02E−5	2.23E−11	1.152E−3
AU coal (production and ocean transport)	3.6E−3	4.482E−09	0.3092
SAf. coal (production and ocean transport)	3.29E−4	3.33E−10	0.04338
SAm. coal (production and ocean transport)	5.45E−4	6.89E−10	0.0779
Domestic coal (open mine and underground mine)			
Coal transport (road)	1.08E−3	1.735E−9	0.019278
Coal transport (rail)	6.82E−3	5.63E−9	0.130
Electricity	6.43E−4	5.04E−11	8.33E−3

Coal stock and infrastructure for mining	5.67E−3	9.34E−10	2.0898
Furnace for gasifier operation	3.42E−6	2.214E−13	9.61E−5
Bed material, catalysts, etc.	1.67E−3	1.692E−10	0.0306
Transport and maintenance of all the vehicle transport infrastructure	1.42E03	2.34E−10	0.0243
Disposal of ash and sewage treatment	4.07E−5	9.05E−13	1.23E−6

Sreejith *et al.*

Sreejith *et al. International Journal of Energy and Environmental Engineering* 2013 4:8, doi:10.1186/2251-6832-4-8

The fertilizer application data given in Table 4 and inventory data provided in 'Inventory data for coconut cultivation' are incorporated. As per the equation, the direct N_2O emissions from the utilization of two nitrogen-containing fertilizers (urea and ammonium sulfate) are estimated as 0.00062 kg CO_2 equivalent for 1 MJ of syngas energy production.

For coal gas life cycle, the major emissions are caused by steam generation for gasification of coal, mining, and oceanic transport of the imported coal. The major contributing greenhouse gas to GWP is CO_2, with contributions of 95.9% and 96.2% of the total for producer gas life cycle and coal gas life cycle, respectively, as shown in Table 12. Power requirement for infrastructure and operation of the gasification plant being the same for these two life cycles, it is inferred that the decisive processes are coal mine operation and coal transport for coal gas life cycle.

Table 12: Major GHG emissions causing global warming

Producer gas life cycle			Coal gas life cycle		
Gas	GWP (kg CO2equivalent)	% of total GWP	Gas	GWP (kg CO2equivalent)	% of total GWP

CO2	0.10473	95.96	CO2	0.128	96.20
CH4	0.00233	2.15	CH4	0.0391	2.98
Dinitrogen monoxide	0.00062	1.52			
CO	0.0000217	0.02			

Sreejith *et al.*

Sreejith *et al. International Journal of Energy and Environmental Engineering* 2013 4:8, doi:10.1186/2251-6832-4-8

The most significant contributions for coconut life cycle are from offsetting Indian electricity generation by coconut palm residues for by-product allocation. This process is favorable (GWP is negative) since fossil fuel-based electricity production is offset by the energy content (16.5 kJ/kg) of the residues, as mentioned in 'Allocation procedure'. In addition to this, allocation of the co-product (coconut oil) is also favorable since it replaces the commercial paraffin oil lubricant.

Two options are suggested for reducing the GHG emissions: (1) increasing the domestic mining capacity of coal, which will reduce the oceanic transport power requirement for importing the coal from other countries, as mentioned in 'Coal gas life cycle'; and (2) increasing the share of renewable and hydroelectric energy source contribution to electricity generation. At present, they are contributing only 16% of the total production capacity, as mentioned in 'Supporting process'. Promotion of these two energy sources, especially renewable bioenergy and small hydropower, shall result in reduction in the emissions of greenhouse gases for both the life cycles.

Ozone Layer Depletion Potential

Ozone layer depletion is a situation of serious concern as it will increase the amount of ultraviolet rays from the sun entering the atmosphere. The emissions of stratosphere ozone layer-depleting gases in the two life cycles are given in Table 13. The major gases responsible for the depletion of the ozone layer is CFC-10 (66.26% of ODP) for producer gas life cycle and halon 1301 (66.11% of ODP) for coal gas life cycle. The results provided in Table 9 show that ODP is reduced by 64.1% in the coconut shell-generated producer gas life cycle. The corresponding process contributions are provided in Tables 10 and 11. The by-product and co-product allocations are responsible for the reduction in ODP for producer gas life cycle, as evident from Table 10. Steam generation, fertilizer application, and irrigation are the significant contributors of ozone-depleting gas emissions for producer gas life cycle. Similarly for coal gas life cycle, in addition to steam production for gasifier operation, oceanic transport of coal (especially Australian coal for its large distance) and rail transport of domestic coal are decisive.

Table 13: Major gas emissions causing ozone layer depletion

Producer gas life cycle			Coal gas life cycle		
Gas	ODP (kg CFC 11 equivalent)	% of total ODP	Gas	ODP (kg CFC 11 equivalent)	% of total ODP
CFC 10	5.87E–9	66.26	Halon 1301	1.58E–8	66.11
Halon 1211	1.59E–9	18.49	CFC 10	5.95E–9	24.89
CFC 114	1.27E–9	14.76	Halon 1211	1.69E–9	7.10
			CFC 114	9.85E–10	1.90

Sreejith *et al.*

Sreejith *et al. International Journal of Energy and Environmental Engineering* 2013 4:8, doi:10.1186/2251-6832-4-8

The major source of ozone-depleting gas emissions in both the cycles is the steam generation. The fossil fuel-based (natural gas and coal) electricity production for steam generation contributes 33% of the total ozone-depleting gas emissions. On analyzing the sub-processes, it is traced that 83% of the ODP of steam generation is attributed to the transport and storage of liquefied natural gas onshore and offshore for utilization in power plants. The liquefaction of natural gas is accomplished by cryogenic treatment, which is responsible for the emission of three gases, halon 1301, CFC 10, and CFC 114, to the atmosphere which deplete the ozone layer in the stratosphere. The remaining 17% is contributed by the utilization of diesel for the transport of hard coal to power plant locations. The diesel production stage (crude oil refining) involves many low-temperature extraction processes utilizing refrigerants, which emit ozone-depleting gases. It may also be noted that the other two major processes in producer gas life cycle are fertilizer production and irrigation, while transport of the coal through large distance (e.g., oceanic transport of the imported coal) is prominent for coal gas life cycle. The share of natural gas in electricity production and fueling (diesel) transport contributes ozone-harmful gases to these three sub-processes also.

Fossil Fuel Depletion Potential or Utilization of Nonrenewable Energy

Syngas produced from gasification of biomass feedstock is often cited as an effective alternative to fossil fuels, thus reducing the dependence on fossil fuel reserves. The utilization of nonrenewable energy or fossil fuel depletion potential (FDP) is expressed in megajoules of energy surplus. The FDP values for the two life cycles are shown in Table 9. The results show that a reduction of 71.5% is possible regarding fossil fuel consumption for coconut shell-produced gas life cycle, compared to coal gas life cycle. This significant reduction is attributed to two reasons: (1) utilization of renewable biomass (coconut shell) instead of fossil fuel (coal) and (2) allocation of coconut residues for offsetting fossil fuel-based power production capacity. The process contributions of both life cycles are given in Tables 10 and 11. Source-wise fossil fuel consumption is represented in Table 14 for both life cycles. The major fossil fuel is coal, with contributions of 84.96% and 77.62% for the life cycles, as shown in the table. For coal gas life cycle, the coal is utilized for electricity production (Indian electricity mix) and also as the feedstock in coal gasifier.

Table 14: Major nonrenewable energy sources causing fossil fuel depletion

Producer gas life cycle			Coal gas life cycle		
Material	MJ energy surplus	% of total FDP	Material	MJ energy surplus	% of total FDP

Coal	1.0867	77.62	Coal	3.56	84.96
Natural gas	0.2113	15.09	Natural gas	0.2416	5.77
Oil	0.04485	3.20	Oil	0.234	5.58
Uranium	0.0533	3.81			

Sreejithet *al.*

Sreejith *et al. International Journal of Energy and Environmental Engineering* 2013 4:8, doi:10.1186/2251-6832-4-8

Contribution of Life Cycle Stages to the Impact Emissions

Based on the impact values provided in Tables 10 and 11, the percentage share of major stages in the two life cycles to the impact emissions is shown in Tables 15 and 16. For producer gas life cycle, the major sources of life cycle emissions are generation of gasification agents (superheated steam and compressed air), production and utilization of synthetic fertilizers, and coconut cultivation and husbandry practices, as shown in Table 15. Utilization of fossil fuel-based electricity is used for the generation of steam and compressed air. This is responsible for the substantial share (38.97%, 32.58%, and 33.98% respectively to GWP, ODP, and FDP) of the stage (generation and supply of gasification agents) to the impacts. Nitrogen-containing fertilizer production and application is another life cycle stage which contributes 34.92%, 44.58%, and 40.52% respectively to GWP, ODP, and FDP. Considering these two factors, some improvement measures to the life cycle emissions are also discussed separately ('Sensitivity analysis').

Table 15: Percentage share of producer gas life cycle stages to GWP, ODP, and FDP

Stage in life cycle	Percentage contribution to the environmental impacts		
	GWP	ODP	FDP

Coconut cultivation and husbandry practices	22.73	18.32	22.25
Fertilizer production and application	34.92	44.58	40.52
Coconut shell production and crushing	0.09	0.09	0.08
Generation and supply of gasification agents	38.97	32.58	33.98
Coconut shell gasification in gasifier including infrastructure development and waste disposal	3.29	4.43	3.17

Sreejith *et al.*

Sreejith *et al. International Journal of Energy and Environmental Engineering* 2013 4:8, doi:10.1186/2251-6832-4-8

Table 16: Percentage share of coal gas life cycle stages to GWP, ODP, and FDP

Stage in life cycle	Percentage contribution to the environmental impacts		
	GWP	ODP	FDP
Imported coal production and transport	4.33	27.62	10.86
Domestic coal production and transport	10.67	34.94	53.63
Generation and supply of gasification agents	76.12	33.76	32.25
Coal gasification in gasifier including infrastructure development and waste disposal	8.88	3.68	3.26

Sreejith *et al.*

Sreejith *et al. International Journal of Energy and Environmental Engineering* 2013 4:8, doi:10.1186/2251-6832-4-8

For coal gas life cycle, the production of gasification agents and production and transport of domestic coal are the major contributors, as shown in Table 16. More than 85% of GWP and FDP is contributed by these two coal gas life cycle stages. Life cycle emissions from domestic coal production and transport dominate that of the corresponding imported coal since 85% of Indian coal is produced by domestic mining from open and underground mines, as mentioned in 'Coal gas life cycle'.

Electricity supply is a supporting process for many processes such as steam generation, fertilizer production, compressed-air production, furnace operation, crushing of coconut shell, infrastructure establishment for coal mining, etc. in the two life cycles. Fossil fuel-based electricity production is the major contributing process to the impacts for the life cycle stages mentioned. Table 17 lists the contribution of electricity production and supply to the emissions of the life cycles stages mentioned in Tables 15 and 16. From Table 17, it is evident that fossil fuel-based Indian electricity production almost completely controls the impacts from the stages coconut shell crushing (C) and gasification agent generation (D) in the producer gas life cycle. Similar is the case with stage 'C1' for the coal gas life cycle. Substantial electricity-consuming equipments such as crushing mill, boiler, and air compressor are responsible for this. It may also be noted that the emissions associated with coal transport dominates in the production and transport stage of the imported (A1) and domestic (B1) coal.

Table 17: Percentage share of electricity production and supply to the emissions of the life cycle stages

	Producer gas life cycle stage					Coal gas life cycle stage			
	A	B	C	D	E	A1	B1	C1	D1
GWP	57.6	26.8	93.8	99.8	1.6	2.1	29.3	93.8	1.5
ODP	61.1	17.1	98.1	99.3	2.3	0.8	7.2	98.1	2.3
FDP	63.1	19.9	99.5	99.7	7.9	0.2	0.03	99.5	7.9

A, coconut cultivation and husbandry practices; B, fertilizer production and application; C, coconut shell production and crushing; D, generation and supply of gasification agents; E, coconut shell gasification in gasifier including infrastructure development and waste disposal; A1,imported coal production and transport; B1,domestic coal production and transport; C1,generation and supply of gasification agents; D1,coal gasification in gasifier including infrastructure development and waste disposal.

Sreejith *et al.*

Sreejith *et al.* *International Journal of Energy and Environmental Engineering* 2013 4:8, doi:10.1186/2251-6832-4-8

In short, it can be estimated from Tables 15 and 17 that the contribution of Indian electricity production and utilization to GWP, ODP, and FDP are 68.8%, 51.4%, and 56.3%, respectively, for the life cycle of producer gas. This is due to the fact that 82% of Indian electricity generation is a combination of coal and natural gas, as mentioned in 'Supporting process'. This fact necessitates switching over from fossil fuel-based electricity to renewable-based green electricity (discussed in 'Sensitivity analysis').

Other Impact Categories

In addition to the discussed impact categories of GWP, ODP, and FDP, IMPACT 2002+ links all types of LCI results to four damage categories: human health, ecosystem quality, climatic change, and resource depletion. Under these damage categories, another eleven impact categories (brief description is provided in the Appendix) are also identified, and the life cycles are compared in terms of these categories also, as shown in Figure 5. The comparison is related to coal gas life cycle which is assumed to be having 100% impact. As evident from the figure, all the impact categories calculated show an edge for producer gas life cycle over coal gas life cycle. Major sources of the emissions responsible for these impacts are represented in Table 18. The difference in the impacts caused by these emissions is due to the contribution of electricity to the grid by coconut palm residues. This is because many of these emissions are resulting from the stack gas from fossil fuel-based power plants. The processes responsible for the emissions of these gases and ions are mentioned briefly for each of these impacts, as given below.

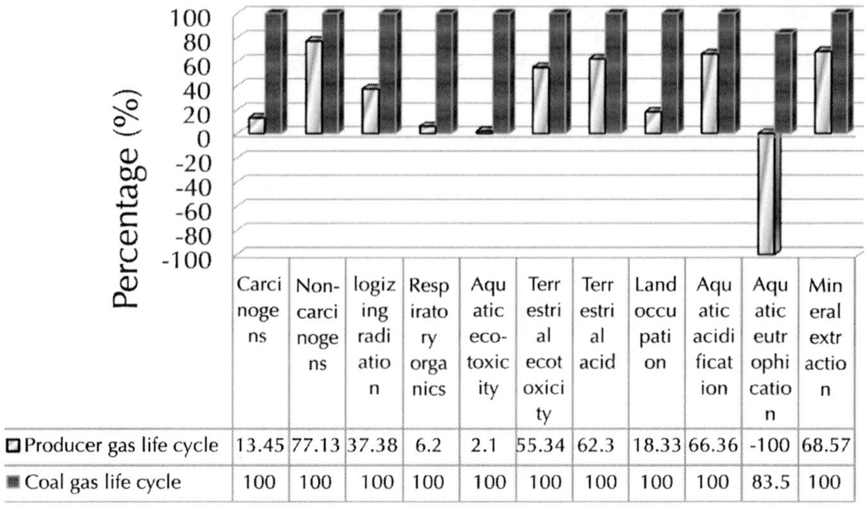

Figure 5: Percentage comparisons of the life cycles based on various emissions.

Table 18: Major emission/substances responsible for various impact categories and their contribution

Emissions	Producer gas life cycle		Coal gas life cycle	
	Emissions/substance	Percentage of total	Emissions/ substance	Percentage of total
Carcinogens	Arsenic ion in water	41.46	Arsenic ion in water	62.87
	Tetrachlorodibenzene-dioxine	27.40	Aromatic hydrocarbons in air	21.41
	Aromatic hydrocarbons in air and water	25.01		
	Polyaromatic hydrocarbons	5.93		
Non-carcinogens (causing human diseases other than cancer)	Arsenic ion in water	73.16	Arsenic ion in water	63.59
	Tetrachlorodibenzene-dioxine	23.51	Selenium	13.06
	Aluminum		Barium	11.42
	Tetrachlorodibenzene-dioxine		5.75	
			3.13	

Ionizing radiation	Radon 222	66.96	Radon 222	76.51
	Carbon-14	26.86	Carbon-14	17.78
	Cesium −137	4.52	Cesium −137	4.97
Respiratory organics	Non-methane volatile organic compounds	26.41	Non-methane volatile organic compounds	85.50
	Xylene	46.08	Methane	3.78
	Pentane	9.47	Xylene	3.52
			Pentane	2.01
Aquatic ecotoxicity	Aluminum	97.03	Aluminum	95.03
	Copper ion	0.47	Copper ion	2.70
Terrestrial ecotoxicity	Aluminum	97.85	Aluminum	69.43
	Zinc	2.11	Copper	14.48
			Zinc	10.93
Terrestrial acid	Nitrogen oxides	76.54	Nitrogen oxides	79.61
	Sulfur dioxide	19.14	Sulfur dioxide	13.11
Land occupation	Occupation forest	40.43	Occupation forest	0.000237
	Occupation industrial area	51.43	Occupation industrial area	0.000133
Aquatic acidification	Sulfur dioxide	62.37	Sulfur dioxide	45.53
	Nitrogen oxides	31.7	Nitrogen oxides	35.24
Aquatic eutrophication	Phosphate	1.38	Phosphate	94.65
	Chemical oxygen demand	95.75	Chemical oxygen demand	4.19
Mineral extraction	Copper	79.94	Copper	79.22
	Aluminum	12.56	Aluminum	12.23

Sreejith *et al.*

Sreejith *et al. International Journal of Energy and Environmental Engineering* 2013 4:8, doi:10.1186/2251-6832-4-8

Carcinogenic and Non-Carcinogenic Emission Potential

The major source is the stack gas emissions from coal and natural gas power plants. Utilization of zeolite for various purposes in these plants makes carcinogenic emissions of elements like arsenic ion and PAH and non-carcinogenic emissions such as selenium. Infrastructure requirement for the power plant operation also contributes (<15%) to these emissions. However, the major emissions for coal gas life cycle are attributed to the tailing of coal in landfill from coal gasification plant (72%).

Respiratory Organics Emissions Potential

A percentage of 91.2 of this impact is contributed by coal power plant emissions of volatile organic compounds and xylene. Emissions from the diesel-fueled vehicles and natural gas power plant emissions are also responsible for less than 8%. For coal gas life cycle, in addition to the 77% contribution from coal power plants and coal gasification plants, the oceanic freighter for coal transport is a critical entity emitting large proportions of methane and pentane.

Ionizing Potential

Emission of species such as radon 222 and carbon 14 by nuclear power plants amounts to 73% of this impact. In the remaining, less than 21% is contributed by the emissions of radon 222 and cesium from coal-fired plants. Uranium enrichment process for fueling the nuclear power plant is also responsible for the emissions of radon 222.

Aquatic Ecotoxicity Potential

The major sources of this impact for both life cycles are the utilization of zeolite powder at natural gas and coal-fired power plants (52%), contamination of metals like Al and Cu in hot water released into the sea by the coal power plant (41%), and utilization of water for various processes in the life cycles (4%).

Terrestrial Ecotoxicity Potential

For producer gas life cycle, 93% of terrestrial ecotoxic emissions and 94.5% of terrestrial acidic emissions are contributed by solid waste emissions to the soil by coal-fired power plants to landfill. In the case of coal gas life cycle, the transport fuel (diesel) emissions are also responsible as its utilization is much larger than that for producer gas life cycle.

Land Occupation

For producer gas life cycle, the occupation of arable land by coal power plant, hydroelectric power plant, and gasification plant share this impact in the ratio of 76%:14%:11%. However, for coal gas life cycle, land for open mines and land for rail and road (for coal transport) are prominent (total 69%), leaving land utilization for power plant contributing less than 20%.

Aquatic Acidification Potential

The presence of sulfur-containing compounds in the stack emissions from coal-fired power plants and the nitrous oxide emissions from fertilizers for coconut cultivation are the major contributors to impact caused by aquatic acidifying emissions.

Aquatic Eutrophication Potential

Utilization of single super phosphate as fertilizer for coconut cultivation emits P_2O_5. For coal gas life cycle, coal tailing, emissions from coal power plant, and crude oil production are the major sources of aquatic eutrophication. In coal gas life cycle, the transport of imported coal by oceanic tanker also is a contributor since the emissions from transport fuel is very high.

Mineral Extraction

Infrastructure for electricity production for each of the power plant and emissions from the utilization of zeolite powder are responsible for the extraction of minerals like Al and Cu. In addition, iron ore from mines are also responsible for the impact caused by mineral extraction for coal gas life cycle.

Energy and Exergy Consumption

The energy and exergy consumptions of the two life cycles in terms of renewable and nonrenewable energy sources are given in Table 19. The reason for the negative values of cumulative energy demand (CED) and cumulative exergy demand (CExD) for renewable biomass source is accounted when allocating the by-product coconut palm residues for electrical energy contribution to the main grid. Exclusion of this by-product increases the CED for producer gas life cycle to 2.3 MJ equivalents. In such case, the renewable fraction in total energy demand is 62.9%, while it is only 2.8% for coal gas life cycle. Substantial improvement in renewable fraction can be achieved by increasing the share of renewable fraction in total power production capacity of the country. To assess this, different scenarios are generated for the share of electricity production, and the results are given in Table 20. As expected, substantial increase in the renewable fraction of the cumulative energy consumption can be achieved by increasing the share of renewable energy in the Indian electric grid.

Table 19: Energy and exergy consumption

Cumulative energy demand (MJ equivalent)			Cumulative exergy demand (MJ equivalent)		
Impact category	Coal gas life cycle	Producer gas life cycle	Impact category	Coal gas life cycle	Producer gas life cycle
Nonrenewable, fossil	4.235	1.37	Nonrenewable, fossil	4.08	1.36
Nonrenewable, nuclear	0.0632	0.0492	Nonrenewable, nuclear	0.127	0.0533
Renewable, kinetic	0.00322	0.00294	Renewable, biomass	0.00607	−1.98
Renewable, solar	0.00254	0.00248	Renewable wind, solar, geothermal	0.00595	0.0056
Renewable, potential	0.11	0.0978	Renewable, water	0.11	0.0978

Nonrenewable, primary	3.87E–7	9.12E–7	
Renewable, biomass	0.00638	–2.08	
Renewable, water	1.8	0.483	
Renewable, metals	0.00406	–0.00644	
Nonrenewable, minerals	0.00167	0.000704	

Sreejith *et al.*

Sreejith *et al. International Journal of Energy and Environmental Engineering* 2013 4:8 doi:10.1186/2251-6832-4-8

Table 20: Renewable fraction of CED for various electricity production scenarios

Share of total electricity production (%)					Renewable fraction of cumulative energy demand (%)	
Coal	Natural gas	Hydroelectric	Nuclear	Renewable	Coal gas life cycle	Producer gas life cycle
70	12	15	2	1	2.8	62.9
60	12	25	2	1	5.2	69.4
56	12	25	2	5	10.7	76.3
56	18	19	2	5	8.5	71.2
35	20	25	0	20	18.9	84.5

Sreejith *et al.*

Sreejith *et al. International Journal of Energy and Environmental Engineering* 2013 4:8, doi:10.1186/2251-6832-4-8

Sensitivity Analysis

Sensitivity analysis is a systematic evaluation process for describing the effect of input variations in a system on the output [6]. The life cycle emissions of the producer gas are broadly divided into five contributors (A to E), as given below.

A. Coconut cultivation and husbandry practices (seedling, bed preparation, tillage, plowing, and irrigation)
B. Fertilizer production and application
C. Coconut shell production and crushing
D. Generation and supply of gasification agents (air and steam)
E. Coconut shell gasification in a gasifier including infrastructure development and waste disposal

Summary of GWP, ODP, and FDP of the life cycle in terms of these broad contributors (A to E) are illustrated in Figure 6 (Table 15 shows the corresponding percentage values). The contributions from by-product and co-product allocation are not represented. The existing fertilization practice (type and schedule) for coconut cultivation in the region and the utilization of gasifying agents for FB gasification of coconut shell are the major contributors to the discussed environmental impacts for the producer gas life cycle. The major reasons attributed to the higher sensitivity of these processes towards GWP are the utilization of fossil fuel-based electricity (68.8%) and the presence of nitrogen content in synthetic fertilizers (25.6%). Out of the 34.92% share of synthetic fertilizer production and utilization (Table 15) to total GWP, 26.8% is contributed by electricity from the grid (Table 17).

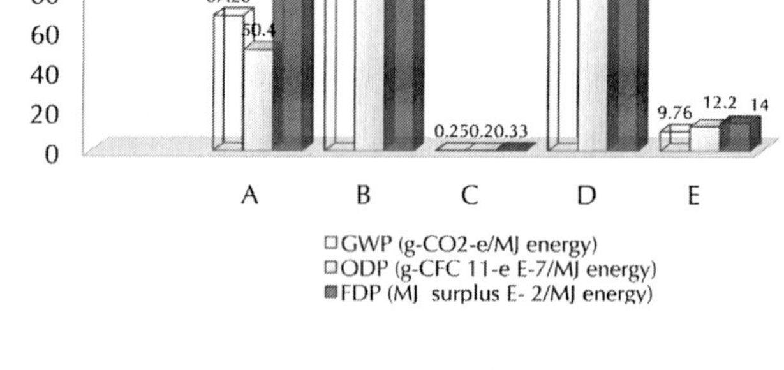

Figure 6: Major contributors to the life cycle impacts of producer gas. A, coconut cultivation and husbandry practices; B, fertilizer production and ap-

plication; C, coconut shell production and crushing; D, generation and supply of gasification agents; E, coconut shell gasification in a gasifier including infrastructure development and waste disposal.

Scenario generation approach, which is widely used [38-40] in LCA studies for suggesting improvement options to environmental impacts, is adopted here to address the GHG emission savings. In the background of the information provided by the data presented in Tables 15 and 17, three scenarios were analyzed to account for their sensitivity on the base case GWP: (1) green electricity, (2) organic farming, and (3) catalytic gasification.

Green Electricity

Complete replacement of fossil fuel-based electricity by renewable energy-based (green) electricity is not feasible in India in the near future, on behalf of the negligible current share (1%) of green electricity. Thus, generation of gasifying agents (air and steam) for gasification is taken as accomplished by energy source contribution in the following ratio: coal 50%, biomass 20%, hydropower 16%, natural gas 12%, and nuclear 2%. This means that the 20% share of fossil fuel is replaced by biomass in the Indian electric grid. A reduction of 11.1% in GHG emissions from the base case (GWP = 108 g CO_2 equivalent) was estimated. The GHG emission savings over coal gas case is 27.7%.

Organic Farming

Application of synthetic fertilizers containing nitrogen emits nitrous oxides. A case in which 50% weight of synthetic fertilizers (ammonium sulfate, urea) replaced by organic fertilizers for coconut cultivation is generated. This results in improvement in GHG emissions by 5.2% from the green electricity-applied base case. The combined GWP improvement over coal gas is 31.5%.

Catalytic Gasification

The lower heating value of syngas generated from steam-assisted AFB gasification of biomass ranges from 6 to 9 MJ/Nm^3[41]. LHV depends on several factors such as design of the gasifier, temperature of steam and air, and particle size of the crushed biomass. In the present study, LHV is taken as 7 MJ/Nm^3. It was reported by Sutton et al. [42] that application of catalysts helps to improve the yield of hydrogen by 20% to 40% (depending on the type of catalyst used), increasing the heating value correspondingly. Assuming a 30% improvement with catalyst application, the volume of syngas to be generated for 1 MJ of energy content (reference functional unit) could be reduced correspondingly. Incorporating this case (when all the three scenarios are implemented at the same time), 43.4% reduction in GHG emissions from the corresponding fossil coal gas case can be achieved. As reported by Singh et al. [39], this reduction in GWP ensures the sustainability of coconut shell-derived producer gas as a biofuel up to 2017, as defined by the EU renewable energy directive [43]. Figure 7 illustrates the GWP improvement accomplished under the scenarios discussed.

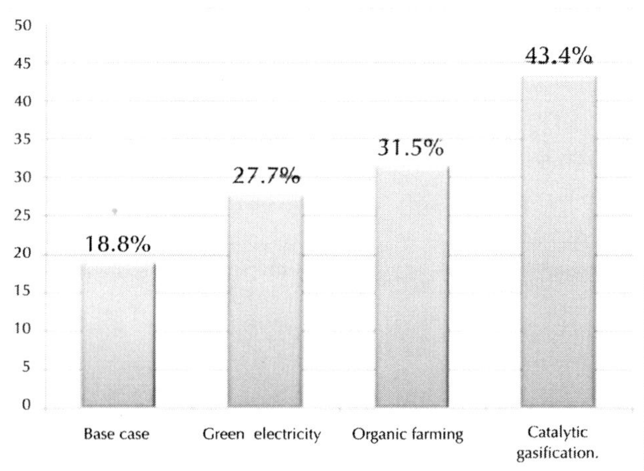

Figure 7: Percentage of GHG emission savings for producer gas life cycle under various scenarios. The scenarios are combined from left to right.

Comparison of the Results with Similar Studies

Several studies reported on the comparative LCA of biofuels and their utilization systems with respect to fossil fuel-based systems. Some of the recent studies reported on LCA of biomass-derived syngas (or its utilization system) and are given in Table 21. Major relevant results drawn from these studies are also mentioned. According to Singh et al. [6], differences in approaches and assumptions taken under consideration in various LCA studies can lead to significant variations in the environmental impacts. It may be noted that direct comparison between the results is not recommended. This is because scaling of LCA balances to geographical scale may not be appropriate due to the variability in topography, soil and climatic conditions, and factors (economical and political) influencing land use [44,45].

Table 21: Some recent studies on LCA of syngas generated by gasification of biomass

Criteria	Siegl et al.[1]	Roedl[14]	Guest[16]	Pa et al.[17]	Puy et al. [19]
Location	Austria	Norway	British Columbia	Barcelona metropolitan area	Sweden
Biomass material	Energy crops, manure, organic residues, wood chips	Forest residues	Wood wastes, wood pellets	Post-consumer wood, forest residues	Municipal solid waste
Reference functional unit	Kilowatt hour electricity generation in power plant which utilizes the syngas	Production of 1 MJ of thermal energy and 1 MJ of electrical energy in CHP plants	1 MJ of heat output	1,000 MJ of energy	1 ton of biomass
Impacts analyzed	GWP, ACD, TE, ET, ODP, ADP, PO, HT	GWP, PO, HT, AD	GWP, combined effect of the impact categories on human health	AD, GWP, ODP, HT, ACD, ET	AD, ACD, ET, GWP, ODP, HT, AE, TE, PO
Reference system	Anaerobic digestion power plant	Various capacity (micro, small, medium) gasification-CHP plants	Natural gas-based district heating system	Comparison between the two biomass life cycles	Sanitary land fill, incineration

Major relevant resultsa	64% to 83% reduction in GWP and 40% to 50% reduction in ODP	84% to 98% contribution of GWP from small- and medium-scale plants (for electricity)	82% and 83% reduction in GWP for wood wastes and wood pellets, respectively	34% reduction in GWP and 66% reduction in ODP for post-consumer wood biomass	3% reduction in GWP and 26% increase in ODP with respect to incineration
		59% to 72% contribution of ODP from small- and medium-scale plants (for electricity)			44% reduction in GWP and 46% reduction in ODP with respect to landfill

[a]Relevant in the context of the present study. AD, abiotic depletion potential; ACD, acidification potential; ET, eutrophication potential; GWP, global warming potential; ODP, ozone layer depletion potential; HT, human toxicity potential; AE, aquatic ecotoxicity potential; TE, terrestrial ecotoxicity potential; PO, photochemical oxidation potential; CHP, combined heat and power.

Sreejith *et al.*

Sreejith *et al. International Journal of Energy and Environmental Engineering* 2013 4:8, doi:10.1186/2251-6832-4-8

Prospect of Coconut Shell in Electrifying India

The purpose of any LCA study is to assess the system in the relevant and selected perspective, and hence, the suggestions and results hold good mostly in the selected system boundary only. The present study is applicable to India, which is the third largest coconut-producing country, behind the Philippines and Malaysia, having an area of about 1.78 million ha under the crop. Annual production is about 7,562 million nuts with an average of 5,295 nuts/ha. The state of Kerala, where the study pertains to, tops in production accounting 39% of the total production in the country. Coconut farming provides employment to nearly 10 million people and makes a contribution of nearly INR 70 billion to gross domestic product. The electricity sector in India had an installed capacity of 205.34 GW as of June 2012, the world's fifth largest [46]. In terms of fuel, coal-fired plants account for 70% (143 GW) of India's installed electricity capacity [47,48]. As per the data from the local agricultural university (as given in 'Life cycle inventory analysis'), 3.1 numbers of husked nuts are required to produce 1 kg of coconut shell. The average calorific value (LHV) of coconut shell ranges from 16.5 to 18 MJ/kg. Thus, based on the above mentioned annual production of nuts, by allocating 100% of the coconut shell for the production of power by converting it into syngas (producer gas), 40.25 to 43.9 GW of power could be produced. This is 19.6% to 21.4% of the total energy demand in the country, offsetting almost 35% to 38% of coal consumption. Careful planning and long-term policies on coconut cultivation encouragement, effective utilization of energy from coconut residues, and coconut shell gasification practice will definitely help to build an effective alternative fuel sector in India.

CONCLUSIONS

Two gaseous fuels (coconut shell-derived producer gas and coal gas) were compared in terms of their environmental impacts by conducting a LCA study with respect to Indian conditions. The environmental impacts are quantified as 18.3% lower GWP, 64.1% lower ODP, and 71.5% lower FDP, indicating better environmental performance of producer gas life cycle. In addition to this, comparisons based on several other impact categories also confirmed the edge of producer gas life cycle. Steam and compressed-air generation for gasifier operation and fertilizer production and application for coconut cultivation are the major contributors to the environmental impacts for producer gas life cycle. Substantial contribution of the by-product allocation is the major factor responsible for lower emissions for the producer gas life cycle. Based on the existing fertilizing practice and the utilization of electricity from the Indian electric grid, substantial GHG emission savings cannot be achieved for producer gas life cycle. However, the green electricity-aided catalytic gasification of coconut shell produced by organic farming can result in 43.4% reduction in GHG emissions from fossil coal gas, meeting the EU renewable energy directive.

AUTHORS' CONTRIBUTIONS

CCS conducted the life cycle assessment and drafted the manuscript. CM conducted the comparison between the life cycles, supervised the work and corrected the manuscript. AP performed the inventory collection, co-supervised the work and helped in drafting the manuscript. All authors read and approved the final manuscript.

REFERENCES

1. Siegl, S, Laaber, M, Holubar, P: Green electricity from biomass, Part 1: environmental impacts of direct life cycle emissions. Waste Biomass Val. 2, 267–284 (2011).
2. Koroneos, C, Dompros, A, Roumbas, G: Hydrogen production via biomass gasification—a life cycle assessment approach. Chem. Eng. Proc. 47, 1261–1268 (2008).
3. Varun, Bhat, IK, Prakash, R: LCA of renewable energy for electricity generation systems—a review. Ren. Sust. Energy. Rev. 13, 1067–1073 (2009).
4. Merola, SS, Tornatore, C, Marchitto, L, Valenteno, G, Corcoine, FE: Experimental investigations of butanol-gasoline blends effects on the combustion process in a SI engine. Int. J. Energy Environ. Eng. 3, 6 (2012).
5. Self, SJ, Reddy, BV, Rosen, MA: Review of underground coal gasification technologies and carbon capture. Int. J. Energy Environ. Eng. 3, 16 (2012).
6. Singh, A, Pant, D, Korres, NE, Nizami, AS, Prasad, S, Murphy, JD: Key issues in life cycle assessment of ethanol production from lignocellulosic biomass: challenges and perspectives. Bioresour Technol. 101, 5003–5012 (2010).
7. ISO: International Standard, ISO, Switzerland (2006)
8. Manish, S, Pillai, IR, Banerjee, R: Sustainability analysis of renewables for climate change mitigation. Energy Sust. Dev. 10(4), 25–36 (2006).
9. Nigam, PS, Singh, A: Production of liquid biofuels from renewable resources. Prog. Energy Comb. Sc. 37, 52–68 (2011).
10. Shie, JL, Chang, CY, Chen, CS, Shaw, DG, Chen, YH, Kuan, WH, Ma, HK: Energy life cycle assessment of rice straw bio-energy derived from potential gasification technologies. Biores. Tech. 102, 6735–6741 (2011).
11. Corti, A, Lombardi, L: Biomass gasification integrated combined cycle with reduced CO2 emissions: performance analysis and life cycle assessment (LCA). Energy. 29, 2109–2212 (2004).
12. Carpentieri, M, Corti, A, Lombardi, L: Life cycle assessment (LCA) of an integrated biomass gasification combined cycle (IBGCC)

with CO_2 removal. Energy Conv. and Mngmnt. 46, 1790–1808 (2005).

13. Rafaschieri, A, Rapaccini, M, Manfrida, G: Life cycle assessment of electricity production from poplar energy crops compared with conventional fossil fuels. Energy Conv. & Mngmnt.40, 1477–1493 (1999).
14. Roedl, A: Production and energetic utilization of wood from short rotation coppice—a life cycle assessment. Int. J. of Life Cycle Assess. 15, 567–578 (2010).
15. Kimming, M, Sundberg, C, Nordberg, A, Baky, A, Bernesson, S, Noren, O, Hansson, PA: Biomass from agriculture in small scale combine heat and power plants – a comparative life cycle assessment. Biomass Bioenergy. 35, 1572–1581 (2011).
16. Guest, G, Bright, RM, Cherubini, F, Michelesen, O, Stromman, AH: Life cycle assessment of biomass-based combined heat and power plants. J. of Ind. Ecology. 15, 908–921 (2011).
17. Pa, A, Bi, XT, Sokhansanj, S: A life cycle evaluation wood pellet gasification for district heating in British Columbia. Biores. Tech. 102, 6167–6177 (2006)
18. Pa, A, Jill, C, Bi, XT, Staffan, M, Sokhansanj, S: Environmental footprints of British Columbia wood pellets from a simplified life cycle analysis. Int. J. of Life Cycle Asses. 17, 220–231 (2012).
19. Puy, N, Rieradevall, J, Bartroli, J: Environmental assessment of post-consumer wood and forest residues gasification: the case study of Barcelona metropolitan area. Biomass Bioenergy. 34, 1457–1465 (2010).
20. Khoo, HH, Tan, RBH, Sagisaka, M: Utilization of woody biomass in Singapore: technological options for carbonization and economic comparison with incineration. Int. J. of Life Cycle Assess. 13, 312–318 (2008).
21. Zaman, AU: Comparative study of municipal solid waste treatment technologies using life cycle assessment method. Int. J. of Env. Sc. Tech. 7, 225–234 (2010)
22. Zhong, ZW, Song, B, Zaki, MBM: Life-cycle assessment of flash pyrolysis of wood waste. J. Clean. Prod. 18, 1177–1183 (2010).
23. Iribarren, D, Peters, JF, Dufour, J: Life cycle assessment of transportation fuels from biomass pyrolysis. Fuel. 97, 812–821 (2012)

24. Bartolozzi, I, Rizzi, F, Frey, M: Comparison between hydrogen and electric vehicles by life cycle assessment: a case study in Tuscany. Italy. App. Energy. 101, 103–111 (2012).
25. Wulf, C, Kaltschmitt, M: Life cycle assessment of hydrogen supply chain with special attention on hydrogen refuelling stations. Int. J. of Hydrogen Energy. 37(21), 16711–16721 (2012).
26. Hawkins, TR, Gausen, OM, Strømman, AH: Environmental impacts of hybrid and electric vehicles—a review. Int. J. of Life Cycle Assess (2012).
27. Balat, M, Kirtay, E: Hydrogen from biomass–present scenario and future prospects. Int. J. of Hydrogen Energy. 35, 7416–7426 (2010).
28. Pollex, A, Ortwein, A, Kaltschmitt, M: Thermo-chemical conversion of solid biofuels. Biomass Conv. and Bioref.. 1, 21–39 (2011)
29. PRO Kerala: Kerala Climate
30. Coconut Development Board:
31. Kerala Agricultural University:
32. Coal India Limited: Coal import export data
33. Central Power Research Institute: Govt. of India
34. Gnansounou, E, Dauriat, A, Villegas, J, Panichelli, L: Life cycle assessment of biofuels: energy and greenhouse balances. Bioresour Technol. 100(21), 4919–4930 (2009).
35. Thottackkad, MN, Perikinalil, RK: Kumarapillai, PN: Experimental evaluation on the tribological properties of coconut oil by the addition of CuO nanoparticles. Int. J. Prec. Eng. Manufact. 13(1), 111–116 (2012).
36. Product Ecology Consultants: Simapro Software
37. United Nations Framework Convention on Climatic Changes: Estimation of direct nitrous oxide emission from nitrogen fertilization
38. Korres, NE, Singh, A, Nizami, AS, Murphy, JD: Is grass biomethane a sustainable transport biofuel?. Biofuels Bioproducts Biorefin. 4, 310–325 (2010).

39. Singh, A, Nizami, AS, Korres, NE, Murphy, JD: The effect of reactor design on the sustainability of grass biomethane. Ren. Sust. Energy Rev. 15, 1567–1574 (2011).
40. Pandyaswargo, AH, Onoda, H, Nagata, K: Energy recovery potential and life cycle impact assessment of municipal solid waste technologies in Asian countries using ELP model. Int. J. Energy Environ. Eng. 3, 28 (2012).
41. Loha, C, Chatterjee, PK, Chattopadhyay, H: Performance of fluidized bed steam gasification of biomass – modeling and experiment. Energy Conv. Manag. 52, 1583–1588 (2011).
42. Sutton, D, Kelleher, B, Ross, JRH: Review of literature on catalysts for biomass gasification. Fuel Proc. Technol. 73, 155–173 (2001).
43. EU Renewable Energy Directive: Directive 2009/28/EC of the European Parliament and of the Council of 23 April 2009 on the promotion of the use of energy from renewable sources and amending and subsequently repealing Directives 2001/77/EC and 2003/30/EC. Official J. Euro. Un. 16, 16–62 (2009)
44. Schmer, MR, Vogel, KP, Mitchell, RB, Perrin, RK: Net energy of cellulosic ethanol from switchgrass. Proc. Natl. Acad. Sci. USA. 105, 464–469 (2008). | PubMed Central Full Text
45. Gurgel, A, Reilly, JM, Paltsev, S: Potential land use implications of a global biofuel industry. J. Agric. Food Ind. Org. 5(2), 1–36 (2007)
46. Central Electricity Authority, Ministry of Power, Government of India: All India region-wise generating installed capacity of power
47. Ministry of Power, Government of India: Power sector at a glance "ALL INDIA"
48. World Coal Institute: The coal resource, a comprehensive overview of coal
49. Product Ecology Consultants, Washington: SimaPro Database Manual
50. IPCC: Fourth assessment report

Chapter 8

Birds, Beasts and Bovines: Three Cases of Pastoralism and Wildlife in the USA

Lynn Huntsinger[1], Nathan F Sayre[2], and JD Wulfhorst[3]

[1]Department of Environmental Science, Policy, and Management, University of California, 130 Mulford Hall, Berkeley, CA, 94720, USA

[2]Department of Geography, University of California, 507 McCone Hall, Berkeley, CA, 94720, USA

[3]Department of Agricultural Economics & Rural Sociology, University of Idaho, Moscow, ID, 83844, USA

ABSTRACT

Background

Pastoralism in the USA began coincidently with the initiation of profound ecological change resulting from colonization in the sixteenth century. Relationships between pastoralism and wildlife conservation in three different contexts of land tenure, environmental legacy, and geography are examined.

Results

On the federal rangelands of the Intermountain West, based on limited scientific information, wildlife policy has been interpreted to require separation of native bighorn sheep from livestock to prevent disease transmission. Ignored are the possible long term and broad scale impacts of removing grazing on the ecosystem and the 'social disturbance' to local communities. In southwestern deserts, the implementation of wildlife policy exemplifies the contradictions between conservation of individuals versus populations, and fire suppression and grazing removal as 'inactions' requiring no review versus grazing and burning as 'actions' requiring regulation and control. In California's Mediterranean rangeland, wildlife policy under the Endangered Species Act is at once a regulatory burden and an opportunity for ranchers. The opportunities result from an evolving recognition that cessation of grazing can harm wildlife.

Conclusions

In all three cases, the environment has changed and is changing due to ecosystem engineering that alters the resources available to wildlife and plant species. Grazing offers potential benefits as a management tool, and pastoralism a means of maintaining un-fragmented landscapes. Yet, absent adequate ecological information, the assumptions of innate conflict between livestock and wildlife, and that cessation of grazing is not an action, as well as the norms of a politically popular yet ecologically unsupportable discourse of restoration, fill in the gaps.

INTRODUCTION

On a global scale, pastoral lands are generally extensive, of low productivity, and of ambiguous ownership. The pastoral regions in the USA are no exception. Traditional forms of food production from pastoral lands, while uniquely suited to the aridity and unpredictability characteristic of rangelands (Behnke et al. 1993), are low in production on a per hectare basis. Lands used for grazing are at best vast, allowing mobility and flexible use of the landscape to accommodate climatic uncertainties (Niamir- Fuller 1998). A large proportion of US arid rangelands in the West are in government ownership and, until recently, were managed largely for livestock grazing, mining, and rural recreation. Transhumant grazing patterns using public deserts and mountains and private watered lands are common (Huntsinger et al. 2010).

Today, alternate uses for the land often seem to offer greater returns than livestock grazing. Mining, cultivation, and nature-based recreation are common competing land uses around the world. Where land is irrigable in the USA, it is almost always farmed and has long been so though technological advances in irrigation methods feed expansion of cultivated lands. Mining is an encroachment on rangelands that is gaining worldwide attention and is a major competitor for US rangelands. Recreation and tourism are growing and provide income to landowners and governments. In the USA, many recreationists have urban roots and may fear or dislike livestock. The expansion of residential areas into US rangelands brings pastoralism under new kinds of scrutiny as new neighbors are drawn to recreate on nearby lands and expect to be shielded from agricultural activities, wildfires, and other natural hazards. In an era of mass extinction, rangelands are looked to as refugia, and 'the marginal lands that were previously the province of pastoralists are increasingly coming into focus as reserves of biodiversity' (Blench 2001). Large areas of US public rangeland have been withdrawn from grazing or had reductions in grazing in order to be managed for wildlife and nature-based recreation. Can mobile pastoralism persist under such conditions?

Wildlife conservation and mobile pastoralism require similar landscape configurations at the landscape scale. They rely on large areas, travel corridors, water supplies, cover, and consumable vegetation. Efforts to protect pastoral landscapes in and of themselves benefit whole suites of species. However, within this framework, at smaller scales, tradeoffs and conflicts occur. As an ecological process, grazing may change vegetation, soils, waterways, and the species profile. Disease transmission between livestock and wildlife may occur. Pastoralists and their activities, including hunting, predator control, fence construction, and water development, directly and indirectly influence wildlife. In the USA, this brings grazing and its accouterments within the scope of the Endangered Species Act and other laws and regulations that require the protection of rare animal species on private land and rare plant and animal species on government land.

METHODS

We focus on three case studies of the relationships between pastoralist production and wildlife conservation in the USA. Because of the institutional setting in the USA, exemplified by the Endangered Species Act and its outgrowths, the push for wildlife conservation offers pastoralists the proverbial double-edged sword: for some, the push for biodiversity is a threat to grazing practices and territory; for others, it may prove the salvation of the pastoralist enterprise as part of a diversification of income streams. Here, we explore the particular configuration of environmental, social, and institutional settings that support these alternative outcomes in the USA by examining wildlife-pastoralist relationships in Mediterranean, southwestern desert, and cold desert steppe regions (Figure 1).

Figure 1: Major vegetation zones of the western USA. The three case studies are in Mediterranean California rangelands, southwestern desert rangelands, and the Intermountain cold desert steppe.

BACKGROUND

Ranching in the Arid West

In the arid lands of the American West, those who graze extensive rangelands with herds of livestock are sedentary, part of the developed economy, and often do not depend on the livestock enterprise for a livelihood (Huntsinger et al. 2010). Yet, ranchers using extensive rangelands share many of the attributes that have been identified as characteristics of traditional pastoralists, including strong identification with livestock husbandry as a way of life, distinct subcultures, a general belief that society is somewhat hostile toward them and their activities, mobility and use of land without secure tenure, a mistrust of governmental authority, reciprocal social relations including a tradition of sharing pastures in times of crisis, and dependence on peers (Huntsinger et al. 2010).

The majority of arid US rangelands are under the ownership and administration of federal, state, or local government as a result of historic processes of land allocation. The US Forest Service controls approximately 78 million ha (US Department of Agriculture Forest Service 2011) of mostly montane lands, and the US Bureau of Land Management controls 106 million ha of mostly desert and steppe (United States Department of the Interior Bureau of Land Management 2011). These lands are also used for recreation, mining, and forestry. Additional millions of hectares are part of Indian and military reservations, and are often grazed by livestock. The better-watered lands are in private holdings often as small parcels located on water and surrounded by government land. Typically, these are used in conjunction with grazing permits on government land in a cyclical annual pattern of using highlands, lowlands, and pasture to provide forage all year. The majority of rangeland livestock producers graze herds of cattle, often of English breeds. Typically, ranchers own 200–300 cows and sell the calves from them each year. A distant and declining second is sheep production.

Surveys of US ranchers have shown them to be strongly interested in wildlife and motivated to undertake 'wildlife habitat improvement' practices (Ferranto et al. 2011; Liffmann et al. 2000; Rowe et al. 2001;

Kreuter et al. 2006). The definition of wildlife may be centered on game animals—exemplified by the comment of one rancher interviewed in California when he said 'I manage for the good wildlife'. In fact, the main purpose of ranch ownership for some is to have a private hunting and fishing area (Gosnell and Travis 2005). Ranchers may charge people to hunt on their land though the animals are the property of the state. This income can be a significant augmentation to income from livestock. Because better-watered lands are often in private ranch ownership, private lands may be the richest wildlife habitat in a region (Maestas et al. 2003).

Wildlife Policy and the Endangered Species Act

In general, native wildlife in the USA whether on private or public land is regulated by the federal and state governments. Government agencies are responsible for managing all wildlife and habitat on public lands. On private lands, due to the historical evolution of wildlife regulation, the details of state regulatory schemes vary from state to state. Prior to passage of the Lacey Act in 1900, which regulated commerce in wildlife, many wildlife specimens were captured live, traded, sold, and considered privately owned property. The offspring of these privately owned animals were also considered private property, which explains the case of the American bison (*Bison bison*), the vast majority of which are owned privately like livestock (Lueck 2002). Many exotic animals imported from abroad prior to regulation and even herds of white-tailed deer (*Odocoileus virginianus*) continue to remain under private ownership in a similar manner. These privately owned animals can often be managed and harvested without the same regulations that govern other native wildlife that are regulated by state or federal government. Game ranching can occur with nonnative species, privately owned, or publicly regulated wildlife with oftentimes less restrictive regulations for privately owned wildlife than government regulated wildlife.

Relations between ranchers and wildlife today are strongly influenced by the Endangered Species Act of 1973. Under the Act, the federal government protects native species designated as endangered or threatened, and the habitats upon which they depend. *Endangered* means a species is in danger of extinction throughout all or a significant portion of its range. *Threatened* means a species is likely to become endangered within the foreseeable future. All species of plants and animals, except pest insects, are eligible for *listing* as endangered or threatened, with over 1,200 species currently listed. The Act makes it unlawful for a person 'to harass, harm, pursue, hunt, shoot, wound, kill, trap, capture, or collect' a listed animal without a permit, referred to as *taking* a listed animal or plant. Through regulations, the term *harm* is defined as 'an act which actually kills or injures wildlife. Such an act may include significant habitat modification or degradation where it actually kills or injures wildlife by significantly impairing essential behavioral patterns, including breeding, feeding, or sheltering'. Listed plants are not protected from take, although it is illegal to collect or maliciously harm them on federal land (United States Department of the Interior Fish and Wildlife Service 2011).

Two thirds of federally listed species have at least some habitat on private land. The federal government has developed an array of tools and incentives to encourage management activities that benefit listed and other at-risk species. *Habitat Conservation Plans*, an option added by a 1982 amendment, may be used by landowners, including private citizens, corporations, tribes, states, and counties, to develop property that is inhabited by listed species. Landowners may receive a permit to take such species incidental to otherwise legal activities, provided they have developed an approved plan. These include an assessment of the likely impacts on the species from the proposed action; the steps that the permit holder will take to avoid, minimize, and mitigate the impacts; and the funding available to carry out the steps. The idea is that such planning may benefit not only landowners but also species by securing and managing important habitat while allowing economic development that is planned to provide species conservation.

Safe Harbor Agreements were initiated in the 1990s to provide regulatory assurance for non-federal landowners who voluntarily aid in the recovery of listed species by improving or maintaining wildlife habitat. An agreement allows such landowners to carry on their usual activities without fear of being punished for take that is incidental to their ordinary management activities.

Conservation banks are lands that are permanently protected and managed as mitigation for the loss elsewhere of listed and other at-risk species and their habitat. Credits are supplied by landowners who enter into a Conservation Bank Agreement with the government that commits them to protect and manage their lands for one or more species. Developers and others who need to mitigate for adverse impacts to those same species may purchase conservation bank credits from willing sellers to meet their mitigation requirements (United States Department of the Interior Fish and Wildlife Service 2011). Use of conservation banks is common in Habitat Conservation Plans, and the plan is supposed to consider the proportion and distribution of lands set aside as conservation banks within the planning area and how that will affect wildlife conservation.

The law requires federal agencies to ensure that actions they authorize, fund, or carry out are not likely to jeopardize the continued existence of any listed species or result in the destruction or adverse modification of designated critical habitat of such species (United States Department of the Interior Fish and Wildlife Service 2011). This means that federal agencies are compelled to protect listed species regardless of other management goals, potentially leading to significant changes in allowed grazing. On private lands, the law generally comes into play when the landowner wants to carry out a regulated management activity or develop the land for other uses. Losses of individuals or habitat must be prevented or mitigated. Many states have created their own set of listed species and similar legislation to go with it, for example, the California Environmental Quality Act includes protection of species listed under state criteria.

On federal lands, there is a body of legislation, policy, and management directives that requires protection of native species in general. The National Forest Management Act of 1976 is now the primary statute governing the administration of national forests. It specifically requires national forests to maintain viable populations of 'native and desired nonnative vertebrate species…well distributed in the planning area.' The secretary of Agriculture's Policy on Fish and Wildlife (9500–4) directs the Forest Service to 'manage habitats for all native and desired nonnative plants, fish and wildlife species to maintain viable populations of each species; identify and recover threatened and endangered plant and animal species and to avoid actions which may cause species to become threatened or endangered.' Forest Service Sensitive Species Policy (FSM 2670.32) calls on national forests to assist states in achieving conservation goals for endemic species, to complete biological evaluations of programs and activities, avoid and minimize impacts to species with viability concerns, analyze significance of adverse effects on populations or habitat, and coordinate with states, the US Fish and Wildlife Service and the National Marine Fisheries Service. Similar policies apply on Bureau of Land Management and other federal lands.

RESULTS AND DISCUSSION

California's Mediterranean Rangelands: Risks and Opportunities

More than 200 years of *ecosystem engineering* have created novel relationships between pastoralists and wildlife in California. The actions of livestock and their owners have created an ecosystem state change that has modulated the flow of resources to a myriad of other species, fitting the classification of *allogenic ecosystem engineering* as defined by Jones et al. (1994). Jones et al. (1994) suggest that using the concept of ecosystem engineering to understand ecosystems is justified when it improves understanding of ecological interactions. In the case of California, using the concept of ecosystem engineering to examine the relationship between pastoralism and wildlife not only enhances understanding of the ecological interactions that shape the ecosystem but also offers critical insights for management and policy.

California's grassland and savanna ecosystems are in a novel and irrevocable state created by colonization and the importation of new animals and plants. This state change is one that follows on state changes engineered by thousands of years of indigenous management. Unlike pastoral regions in much of the world, livestock grazing has been part of California ecosystems for less than 300 years. Spanish importation of cattle and sheep into California began in 1769. California's Native Americans did not use domesticated grazing animals, and the flora and fauna they managed, largely with fire, were for the most part isolated from the rest of the world for millennia. Ecosystems were, thus, suddenly and heavily influenced by multiple post-contact impacts, including pastoralism, plowing, changes in fire regimes, reduction of native grazers like elk and antelope, and the introduction of aggressive species from the other Mediterranean regions of the world.

The native grasslands of the state have been invaded on a massive scale by aggressive grasses and broadleaves from the 'fertile crescent', plants that have survived and flourished despite thousands of years of widespread cultivation and livestock grazing. The role played by livestock grazing in the success of this invasion, relative to all the other factors that accompanied colonization, is debated (Bartolome and Jackson 2007); however, what is clear is that the grasslands have been profoundly changed by annual grass species that have largely supplanted native species. Pastoralism and farming had a definitive role in that seeds were carried in and on stock and in feed and crop seeds throughout California. Livestock grazing in some areas reduces the presence of shrubs and, by removing biomass, reduces the spread and occurrence of wild and deliberate fires. Grazing also creates soil disturbance, changes mineral cycling through manure deposition, and may favor one perennial species over another, which could have had an influence on some of the native perennial flora. These factors all contributed to the state change. Pastoralism brought over and, then, helped increase pastoralism-adapted flora from the Mediterranean Basin.

The actions of pastoralists have also created ecosystem change particularly through the engineering of water. Irrigation ditches, irrigation itself, and the construction of ponds are prominent examples of engineering associated with pastoralism. Water has been removed from some areas and transferred to others. Water from creeks that would otherwise dry up in summer is stored in ponds that last farther into the dry season, extending the grazing period for livestock. Ditches and pipes transfer water to ponds and water troughs, or to irrigated pastures. Cultivation is also associated with pastoralism to some extent; however, as cultivation usurps the ecosystem and changes it to an entirely anthropogenic state that is broadly recognized as such, it is not included in this analysis. Instead, we focus on the quasi-natural rangelands that are commonly subject to restoration discourse.

Environmentalists, scientists, and natural resource managers often perceive California rangeland ecosystems as in a condition that represents a deviation from a recoverable 'natural' state. Frequent use of terms like 'degraded' reflects a normative view of ecosystem change that is based on the idea that a former 'pristine' state has been shifted in a linearly negative way to a 'lower' condition and remains in that lower state as long as the driver of change remains, returning to a higher state once the driver is removed. This way of seeing the ecosystem has been transformed into management practice, policy, and scientific inquiry that is part of a powerful discourse of 'restoring' California ecosystems by returning ecosystems to native flora. Unfortunately, while the engineers—livestock and pastoralists—may be and are being removed from the system, the engineered artifacts in the form of the changed grassland have an extremely long half-life—well beyond any anticipated time horizon.

The Default Prescription

An assumption of the restoration discourse has been that part of the restoration process must be reduction or elimination of livestock grazing and pastoralism. A movement to restrict or eliminate livestock grazing from sensitive areas and from all government lands has gained considerable momentum in the USA. This view can be linked to a classical Clementian succession model of vegetation change where deviation from the ideal end state or 'climax' can be reversed by the removal of the driver or 'disturbance' that caused the deviation. Livestock grazing has, as a result, been treated as an unnatural disturbance for decades. In line with this view, the removal of 'unnatural' livestock grazing has long been a default step on the path to managing for biodiversity conservation.

Decades of research has been unable to connect reduction or elimination of grazing to a significant revival of native species in California (Huntsinger et al. 2007). Returning to the ecosystem engineering concept, what took place is a state change from a grassland of native to introduced flora. This state change is in large part permanent: practically speaking, the nonnative annuals cannot be removed (Heady 19581977). This shift had profound implications for the fauna that depended on the previous flora. As Jones et al. (1994) point out, such changes benefit some species and disadvantage others. The resources and habitat available to the native flora and fauna changed. The species found in California today are those that have successfully and rapidly coped with or benefited from this state change.

The ecologists, managers, and environmentalists of today, then, derive their understanding of California's flora and fauna from the existing state but generally without acknowledging or understanding its engineered origins. Removal of pastoralism does not stimulate a return to a pristine state nor does it maintain the existing second state but, instead, creates an entirely new third state where the artifacts of engineering persist, but the engineering agent is removed. Again, removal causes a modulation of resource flows, advantaging some species and disadvantaging others. Some of the species that have flourished in the engineered second state environment are lost as this new third state develops, and resource streams are altered. Ironically, studies are accumulating that show that a number of the state's rare species are among those disadvantaged by this shift to a new third ecosystem state (Barry 2011).

Another legacy of Spanish, Mexican, and American colonization is that California has a larger proportion of private landownership than is characteristic of much of the western USA. Half of the state is privately owned, and pastoralists, referred to as ranchers or range sheep producers, own an estimated 9 million ha out of the approximately 18.5 million ha of private land in the state (California Rangeland Trust 2011), mostly in the Mediterranean zone. However, there is growing pressure on even private landowners to manage for restoration, and as private and public lands are acquired for or shifted to conservation purposes, they become subject to restoration goals that include removal or restrictions on pastoralism. On such lands, the removal of grazing and water engineering can have profound and unanticipated impact on wildlife and plant species. As this has become more understood, new opportunities for pastoralists have opened up at least partly as a result of the Endangered Species Act. The following cases illustrate these points. The first, that of the California tiger salamander in the San Francisco Bay Area, is selected because its dependence on grazing and pastoralism is now widely recognized, and institutional mechanisms have been developed to create pastoralist-friendly conservation initiatives. The second, the case of the California Black Rail, is a newly unfolding saga of pastoralism and a rare species where the relationship of the species with grazing may prove more complex.

The California Tiger Salamander: a Case in Point

California has a disproportionately high percentage of threatened or endangered species compared to other states, and the majority of these occur on private lands (Scott et al. 1995; Dobson et al. 1997). One of these is the California tiger salamander (*Ambystoma californiense*). In the San Francisco Bay Area, most native habitat has been eliminated by urban expansion. Of the 155 California tiger salamander locality records in the eastern San Francisco Bay Area (Alameda and Contra Costa counties) where the wetland type was identified, 85% (131 sites) were located in stock ponds (United States Department of the Interior Fish and Wildlife 2004). Stockponds have been created over the last 100 years by developing springs or blocking small waterways, creating dams that hold water into the dry season (Figure 2).

Figure 2: A stockpond in the San Francisco Bay Area. Cattle have grazed off the vegetation at the edge of the pond and muddied the water. Both conditions are now believed benefit of the endangered California tiger salamander. When bullfrogs are absent, stockponds are also crucial habitat for the threatened red-legged frog (*Rana draytonii*) (United States Department of the Interior Fish and Wildlife Service 2006).

The presence of an endangered or threatened species on a rancher's land can lead to fear and apprehension on the part of the landowner. If the landowner chooses to develop the land, the species and habitat must be protected or the lost habitat must be mitigated under the Endangered Species Act. Obviously, this increases the costs of development and reduces the potential sale price of the land. In one famous case in California, environmental consultants inventorying species for a developer were caught secretly moving red-legged frogs to another location to reduce the costs of compliance with the Endangered Species Act (Lee 2001).

Even without concerns about the value of the land, the landowner may find management activities limited by the Act. In a famous case in timberlands, landowners preferred to eliminate potential habitat for an endangered bird before it could be identified on their land so that timber harvest practices would not be restricted (Lueck and Michael 2003). Enforcement of this Act and California species protection regulations is in part responsible for the fact that the US Fish and Wildlife Service and the California Department of Fish and Game are the conservation agencies that forest and rangeland landowners trust the least (Ferranto et al. 2012). Safe Harbor Agreements that allow landowners complying with certain requirements to carry on their usual activities without fear of being punished for take are an attempt to reduce the concerns of landowners. In the case of California ranchers and the tiger salamander, a special exemption was created for routine ranch practices (United States Department of the Interior Fish and Wildlife 2004), at least partly in recognition of the fact that without ranches, there would be no habitat at all.

However, the Act also opens up the possibility of selling conservation bank credits from ranch lands, as one way to 'mitigate' is to purchase or create habitat for the endangered or threatened species elsewhere. Some ranchers have sold 'mitigation easements' to earn some extra funds to support their ranching. An easement is a permanent change in the property title that limits or controls the use of a geographical area. While government wildlife officials are prohibited from entering a private landowner's land without permission, to participate in mitigation, or some conservation cost-share programs, a landowner must allow the species and its habitat to be verified on their land. At the California Society for Ecological Restoration Conference, a rancher related the following about going with a wildlife official to examine his stock pond for the presence of tiger salamanders:

We walked out to the pond and I saw it with fresh eyes. It was denuded, and muddy, and looked just terrible. I thought to myself, 'this is not going to be good.' However, when the wildlife official put his net in the pond, it came out full of salamander larvae. The biologist said he had never seen so many larvae before.

Since these events, research and common experience have shown that tiger salamanders seem to prosper in grazed ponds. A survey of grazed and un-grazed ponds found them more common when vegetation was grazed and water was muddy (DiDonato 2007). Salamanders may be able to move across the grassland when the grass is shorter, predatory insects cannot hide on emergent vegetation when it is grazed, and predators find it harder to target the amphibians when the water is not clear. In fact, salamanders have been found to disappear from fenced and restored ponds. Yet, grazed ponds do not conform to the classic image of a natural, wildlife-friendly pond. Ranchers selling conservation bank credits or mitigation easements on their lands are also often asked to restrict or eliminate grazing on the mitigation areas. One rancher related the following to students on a field trip in 2010:

I had serious problems paying the inheritance taxes on my land. I sold an easement for the protection of a stock pond I have that has salamanders in it. This pond was built by my grandfather, and was excellent for watering cattle. After I sold the easement, the representative from the wildlife agency came out to have a look. He said, 'well, let's talk about where to put the fence to keep the livestock out'. Now this pond had been grazed for decades, and salamanders were common.

After some struggle and various intercessions, the landowner was able to convince the agency to allow him to continue to graze the land. However, once designated as endangered species habitat, activities like pond maintenance can become a major struggle with bureaucracy. Ponds need to be cleared of sediment every so often, and dams need to be repaired. There are cost share funds available from the federal and state governments for improving wildlife habitat or environmental quality. Repair projects generally require five or six environmental permits, and the supervision of a wildlife biologist as part of a lengthy, expensive, and confusing process that brings the ranch under the scrutiny of a regulatory agency. Ranchers tell stories of long delays and, from their perspective, illogical and costly decisions. Some ranchers also fear that they could face new Endangered Species Act restrictions if they voluntarily improve habitat for endangered species. Evidence is accumulating that grazing is beneficial for many species of concern, yet the norms of the wildlife agencies in general run to the opposite view and are slow to change.

In response to these concerns, in one Bay Area county, the Resource Conservation District and Natural Resource Conservation Service, which administer cost-share funds from the government and provide technical advice to ranchers and farmers, developed a program to help ranchers cope with the paperwork and hassles in putting together funding for wildlife-sensitive pond maintenance and for understanding compliance requirements. This program, termed the 'Wildlife Friendly Pond' program, helps ranchers find funds from several sources so that as much as 90% of the costs, often tens of thousands of dollars, can be covered (Alameda County 2011; Environmental Defense Fund 2007). The district helps coordinate permits, acquire Safe Harbor Agreements that protect the ability to use the pond for livestock, and find mitigation and conservation easement opportunities.

A common comment of ranchers is that if society would just pay them directly to raise endangered species ranchers could make them plentiful. Cost shares and conservation banking can bring much needed funding to ranchers who can navigate the paperwork and bureaucracy, and who have an endangered or threatened species on their land that they can manage for and still maintain operations. As a rancher with a tiger salamander easement said at a meeting of the Central Coast Rangeland Coalition in 2011, 'These salamanders are the most valuable livestock I've ever raised'. In the USA, this is one of the forms of 'payment for ecosystem services' that is extant though seldom acknowledged as such. However, these incentives depend on recognition that grazing can be benign and even beneficial, while removing grazing can change the habitat in harmful ways. There is always an element of risk to a rancher when becoming involved in endangered species protection because of the large scale politics attached to federal and state legislation.

The California Black Rail: Interdependence and Tradeoffs

The California Black Rail (*Laterallus jamaicensis cotuniculus*) is a secretive bird that is listed by the state of California as threatened and, for a long time, was thought to remain only in limited coastal marsh areas. In 1994, it was found for the first time in small shallow wetlands scattered through the northeastern California Sierra Nevada foothills. These lands are mostly ranch lands interspersed with expanding patches of residential development. About 89% of its habitat in this region was found to be associated with irrigation ditches that provide water to ranches and farms, with 63% of the habitat on private lands (Richmond et al. 2010). Irrigation ditches and stock ponds that feed livestock watering facilities or irrigate pastures create wetlands where water seeps into the grasslands as does the runoff from irrigated pasture. It is clear that the fate of ranching and the species are intertwined.

Research suggests that grazing may need to be constrained in or excluded from these small natural and artificial wetlands (Richmond et al. 2010), especially during the breeding season and in sites that are not fed by irrigation water (Richmond et al. 2012). The Black Rail lives entirely within thick stands of herbaceous wetland plants and is found in very shallow water (≤3 cm). It will vacate sites that are grazed too intensively. Thus, while the relationship between the bird and the ranching landscape is clearly synergistic, there are potential tradeoffs between livestock grazing and bird use. For individual wetlands, removal of grazing has both long- and short-term results. It may prevent consumption of wetland plants needed to sustain the bird within a particular year, but exclusion of grazing over decades may also lead to vegetation change that can alter the habitat due to invasion of shrubs, vines, and other woody species into the wetland, creating a state three ecosystem. In one extreme case, in an effort to 'restore' the wetlands into conforming with a normative notion of wildlife habitat, one wildlife agency planted trees in some of the wetlands on government land. These are now reaching a size where they shade out the grasses and respire the water that forms the bird's habitat. On a field trip to see Black Rails, one of the well-meaning people who helped plant the trees commented, 'What were we thinking? Now we are going to have to go back in and cut those trees down'.

Researchers studying Rails find it a challenge to obtain the cooperation of landowners, yet they need landowner permission to enter private land for surveys. Landowners fear possible consequences from the Endangered Species Act even though the bird is not federally listed. The longer-term challenge is to help ranchers see the bird as more of an opportunity than a burden as some of the ranchers in the previous case now see the tiger salamander. Unfortunately, as is typical for agricultural operations in California, there are also conflicts among the many regulations and policy initiatives that affect natural resources on private lands. For example, in this area, there is a push to conserve water with local government programs encouraging cement lining of irrigation ditches and fixing 'leaky' irrigation systems. The outcome could be disastrous for the Rails.

One of the counties where the Rails are found is currently developing a Habitat Conservation Plan for the portion of the county that includes all the potential Rail habitat. Still undergoing review, the plan acknowledges the need for protection of large contiguous areas and corridors for protection of some species but does not mention the need to do anything to conserve ranching. While several of the endangered species described in the plan are observed to do better with livestock grazing, where the livestock are going to come from in the long term is not considered. Instead, the draft recommendation is that large areas of land be acquired and managed by the government to conserve wildlife. Even when convinced that livestock grazing is essential for maintaining species abundance, managers and planners often fail to consider the ranches or people that produce stock. As in many parts of the world, government and nongovernmental organizations hire 'professional land managers' or 'environmental consultants' that do not know the rancher's perspective, respect rancher knowledge, or understand the imperatives of the pastoralist operation. They may ask ranchers to graze for a month here or a month there at limited times, making an economic enterprise infeasible. As one rancher commented at a local symposium in 2001:

They don't realize that you cannot just put the cattle on a shelf when you don't need them...they have to go somewhere. Now I am spending money and contributing to carbon emissions by trucking my animals long distances from season to season, because government managers don't want them here in the summer. Is that green? "

Rare Species and Livestock Grazing

Over the last two decades, as grazing has been removed from more areas, examples of cases where removal of grazing has had the opposite of the intended effect, have accumulated (Barry 2011). These include endangered butterflies (Weiss 1999), Burrowing Owls (Nuzum 2005), insects (Dennis et al. 1997), kit fox (Balestreri 1981), kangaroo rats (Kelt et al. 2005; United States Department of the Interior Fish and Wildlife Service 1997), flowers (Barry 2011), and a host of rare flora and fauna associated with vernal pools (Marty 2005Pyke and Marty 2005). In many cases, the reason grazing is a positive force is because grazing helps control the nonnative grasses that otherwise overwhelm many species or change habitat conditions, as in a state three ecosystem.

In addition to benefits in the pasture, at the landscape scale, pastoralism and wildlife conservation have considerable synergy. Pastoralists own much of California's best wildlife habitat and maintain it in a quasi-natural state for rangeland grazing. The imperatives of pastoralism mean large areas of land and livestock mobility, a pattern of land use that is compatible with many wildlife species. When ranchers sell their lands for development, that habitat is lost. When it goes into park status and grazing is removed, it converts to a new state, putting species at risk. Conservationists have become concerned, calling for the support of ranches as a means of conserving open land. Ranchers fear that this might lead to limits on their ability to sell their land when they need to, but they also see opportunities for marketing 'conservation easements' that provide cash payment to ranchers in exchange for giving up the right to develop the land. Nongovernmental entities that broker the sale of conservation easements, such as The Nature Conservancy and the California Rangeland Trust, report that there are more ranchers interested in selling an easement than there are funds to buy them.

Coming up against the norms of environmentalism and the discourse of restoration, the effort to conserve habitat by conserving ranches runs into at least two kinds of trouble related to scale. On the one hand, those interested in conserving ranches to conserve the landscape may want more restrictions on grazing in the pasture, perhaps increasing costs to the point where the ability of the rancher to earn an income is compromised. On the other hand, those demanding the grazing of pastures to maintain wildlife and plant habitat may fail to consider that this also requires conserving ranches as viable enterprises, including attendant mobility and year round grazing opportunities, as pastures persist only as part of ranches.

Although pastoralism had an important role in transporting and facilitating the spread of nonnatives in California, ironically, the success of the nonnatives now makes livestock grazing a part of their management. The ecosystem has been changed irrevocably along with the management options. The environment has been 'engineered' (Jones et al. 1994) by the people that have constructed ponds and ranches and by the livestock themselves. The role of livestock in maintaining biodiversity in Mediterranean ecosystems where grazing has been a land use for thousands of years is acknowledged (Perevolotsky and Seligman 1998), but finding this in a system that has been influenced by livestock for such a short time is unexpected and has both positive and negative consequences for pastoralists.

The Southwestern Desert: a Legacy of Change

As in California, the rangelands of the southwestern USA evolved without livestock until the arrival of Spanish explorers. The region is in the Basin and Range Province characterized by isolated north-south trending mountain ranges separated by broad alluvial valleys with strong elevational and topographic gradients of temperature, precipitation, and vegetation (Figure 3). Grasses dominated the valleys above roughly 900 m, grading into oak savannas, pinon-juniper woodlands, and coniferous forests in the mountains. Bimodal and highly variable precipitation, varied topography, and the biogeographic location of the region—at the interface of the Great Plains, Rocky Mountains, Sierra Madre Occidental, and the Chihuahuan and Sonoran deserts—endowed the southwest with the highest known levels of biological diversity in the USA (Brown and Kodric-Brown 1996).

Figure 3: Cattle graze on the southwestern desert in Arizona.Mesquite shrubs are scatted through the grassland.

Unlike in California, however, the arrival of European settlers, livestock, and associated organisms did not quickly rearrange the southwest's native vegetation. Scholars believe that the first cattle entered the region with the expedition of Francisco Vasquez de Coronado in 1540–1542, and that some of these animals escaped and became feral, leaving descendents that were encountered by Anglo-American explorers and settlers in the 1800s. In New Mexico and Arizona, Native Americans acquired livestock from Spanish settlers and incorporated livestock husbandry into their cultures (Bailey 1980), but environmental historians have found no evidence that these cattle caused widespread or lasting impacts on the landscape (Bahre 1991).

Several factors probably contributed to this. Most importantly, the region's climate differed significantly from most of Europe such that newly arriving plants did not find suitable conditions for rapid establishment and spread. Additionally, both crop agriculture and pastoralism were constrained by limited surface water supplies, periodic droughts, and extreme temperatures, as well as prolonged resistance from the Apache Indians who tenaciously defended large parts of the region well into the 1870s. Moreover, the desert grasslands had evolved to withstand disturbance by native grazers and, especially, by fire. Bison were present in the areas closest to the Great Plains (present day New Mexico and western Texas) and probably ranged in limited numbers into what is now southeastern Arizona. Native perennial grasses such as blue grama (*Bouteloua gracilis*) and black grama (*B. eriopoda*) maintained their dominance over shrub species such as mesquite (*Prosopis* spp.) with the regular assistance of fire, set both by native peoples and by lightning, which is abundant at just the time of year when fires are most likely to spread (in early and mid-summer, after the long spring dry period). As best can be determined, fires recurred in the grasslands at intervals of 8–12 years, setting back shrubs and favoring grasses (Kaib et al. 1996).

Dramatic environmental change occurred beginning in the 1870s with the final conquest of the Apaches, the discovery of massive mineral deposits (especially silver and copper), the arrival of the railroad, and huge influxes of capital from the northeastern USA and Great Britain, much of it underwriting livestock imported from surrounding regions (Dobyns 1981, Bahre 1991). Wetter periods (often associated with *El Nino* years (Swetnam and Betancourt 1998)) produced bountiful forage and encouraged rapid growth in herd sizes which overwhelmed the grasslands during subsequent droughts. The period from 1880 to 1910 encompassed several of these climatic cycles, including severe droughts in 1891–1893 and 1898–1904, when vast areas were grazed bare of vegetation (Bahre and Shelton 1996, Sayre 1999). Heavy rainfall events resulted in arroyo formation throughout the region, permanently altering the hydrology of nearly every major valley (Cooke and Reeves 1976).

This history is profoundly important to issues of wildlife and pastoralism in the Southwest up to the present day, but not in obvious or straightforward ways. Wildlife populations were undoubtedly affected by the environmental changes of the late nineteenth century but with one partial exception, the Masked Bobwhite Quail (Sayre 2002), none were extirpated by grazing impacts. Large predators were subsequently hunted to regional extinction, including the jaguar, Mexican gray wolf and grizzly bear, and some important small mammals, such as the black-tailed prairie dog, were persecuted as pests and dramatically reduced in range and numbers (Oakes 2000). Hunting pressure and habitat changes also reduced populations of game species such as deer, pronghorn, and bighorn sheep. The impacts of regional environmental change on wildlife were otherwise indirect, incomplete, and of difficult attribution, particularly with regard to livestock.

What matters, instead, are the legacy effects of the cattle boom period both on the landscape and in relation to wildlife regulations, especially the Endangered Species Act. Human engineering of southwestern landscapes responded to the crisis in numerous ways, moderating the intensity of grazing impacts but also facilitating their areal expansion through the construction of artificial water sources (including wells, pipelines, and earthen tanks to capture run-off) and fences. In Arizona, a state typical of the region, 74% of the land is federally administered, and 14% is administered as state land. Policy reforms on government rangelands, implemented after 1905 for national forests, 1912 for state lands in Arizona and New Mexico, and 1934 for remaining public domain lands (now administered by the Bureau of Land Management), imposed controls on herd sizes which declined steadily during the twentieth century to less than 25% of peak levels in the 1890s (Sayre 2002). Meanwhile, the landscape responded in ways that were unexpected and poorly understood at the time. Arroyo formation lowered water tables, desiccating former floodplains and converting marshes (known regionally as *ciénegas*) and grassy meadows of sacaton (*Sporobolus wrightii*) into woody corridors (of cottonwood and willow in wetter arroyo bottoms and mesquite forests on former floodplain terraces) (Turner et al. 2003). Miles of perennial surface flow were lost to higher run-off rates and groundwater depletion.

Most important for the region's rangelands as a whole was the interruption of the evolutionary fire regime, with widespread fires largely disappearing after 1893 (Swetnam et al. 1999). This was explicit government policy spearheaded by the Forest Service and applied throughout the region. Livestock grazing was probably even more important than direct fire suppression because it reduced standing fine fuel loads and, thereby, inhibited both ignition and spread across vast areas. Stocking rates varied much less than precipitation. In the Chihuahuan desert, annual rainfall is less than 50% of the long-term average in one year out of two (Herbel and Gibbens 1996). Thus, grazing impacts were most severe during droughts—when fires would have been most likely, and when they had previously been common—effectively ensuring that fires ceased in grazed grassland sites. As artificial waters were installed, more and more of the landscape fell into this category. By the 1930s, shrubs had gotten the upper hand over grasses throughout the region. Because mesquites can withstand fire by resprouting from root crowns once they exceed 1 cm in diameter, even 10–20 years without fire could set woody encroachment in motion for decades to come, gradually reducing grass production and, in a positive feedback, inhibiting fire further (Sayre 2003).

As previously mentioned, both environmental policy and public perception in the USA reflect widespread belief in a singular 'natural' equilibrium state, enshrined in Clementsian ecological theory as the climax to which ecosystems will return if disturbances are removed. Southwestern desert grasslands have proved to be a prime counter-example, and much of the scientific work of discovering this and defining alternative ecological models has been based on research in the region (Westoby 1980). Woody encroachment has occurred even in sites where livestock were permanently removed in the first decades of the twentieth century, and by the 1940s, range scientists were developing mechanical (and later chemical) methods of combating mesquites and other shrubs, effectively abandoning the idea that stocking reductions alone could result in grassland restoration (Sayre 2010).

If restoration of original conditions does not happen naturally, however, this has not prevented the implementation of regulatory policies that tacitly presuppose the contrary. For as long as the Endangered Species Act has been in effect, the listing of species has been attended by calls for the reduction or elimination of livestock grazing on southwestern rangelands. Until recently, federal agencies, environmental NGOs, and many scientists adhered to a common model for endangered species conservation in the region: find the best habitat and protect it by removing human commercial activities which almost always includes livestock grazing. Examples abound, especially among birds (e.g., Masked Bobwhite Quail, Southwestern Willow Flycatcher, and Cactus Ferruginous Pygmy Owl) and fish (e.g., Gila topminnow, Gila chub, and Yaqui chub).

The conventional model is reinforced by legal bureaucratic procedures that obscure the historical and ecological dynamics described above. When a species is listed as threatened or endangered, the Fish and Wildlife Service is obligated to delineate its historical range based heavily on documentation of specimens collected and cataloged for taxonomic purposes. This area becomes the *de facto* baseline for evaluating present distribution and potential critical habitat regardless of whether the area is or could be viable habitat today. Similarly, listing involves specifying known threats to the species (or populations thereof), and livestock grazing is invariably included, if only because of the widespread impacts of the late nineteenth century and the subsequent changes entrained therefrom—regardless of whether current, actual livestock grazing is known to be causing harm. Given that livestock grazing has been widespread and ongoing in the region for well over a century, moreover, it would be reasonable to ask how a listed species could have persisted on grazed rangelands at all if it were fundamentally incompatible with livestock.

In other words, if a rare wildlife species is found on a ranch today, it would be just as logical to attribute a positive role to livestock grazing as a negative one if indeed it plays any role at all. In a few cases, similar to those described from California, this possibility has in fact been recognized and codified by the Fish and Wildlife Service. The Chiricahua leopard frog (*Rana chiricahuensis*), listed as threatened in 2002, has declined with the loss of natural perennial surface waters and the spread of predating bullfrogs. It persists primarily in refugia it has found in ranch stock tanks which provide perennial habitat discontinuous from bullfrog-invaded streams. The Fish and Wildlife Service, therefore, chose to exempt stock tank maintenance activities (such as removing accumulated sediment which is done with bulldozers) from the definition of take for the leopard frog (United States Department of the Interior Fish and Wildlife Service 2002). One Habitat Conservation Plan and three Safe Harbor Agreements are now in effect for various places in the Southwest to ensure that ranch-related activities do not run afoul of endangered species protections for the leopard frog (United States Department of the Interior Fish and Wildlife Service 2002).

More generally, there is growing recognition within the Fish and Wildlife Service and sectors of the environmental community at large that normal, ongoing ranching activities today are benign with respect to most listed species. When the Malpai Borderlands Group, a coalition of ranchers in extreme southeastern Arizona and southwestern New Mexico, conferred with Fish and Wildlife Service and other agency scientists to craft a Habitat Conservation Plan, for example, they reviewed the scientific literature for more than 50 rare species (including ones already listed and others that might become listed) and evaluated a suite of ranching practices (from livestock grazing to stock tank maintenance, fence and waterline construction, and prescribed and wild fire) in relation to each species. They could find only one scenario that might pose a serious risk of constituting take under the Act: if a large prescribed fire were followed by severe rainfall that resulted in sediment deposition sufficient to impact listed fish species downstream of the fire. They subsequently formalized a plan covering 19 species (nine of them listed under the ESA) and providing regulatory protection for various range and livestock management activities. Significantly, the plan does *not* cover livestock grazing in and of itself—that is, herbivory or the consumption of vegetation by livestock—because the probability of take for any covered species was judged to be too remote to justify its inclusion (Lehman 2007).

It is important to understand that this shift in perspective regarding pastoralism and wildlife in the southwestern USA has occurred against a larger backdrop of land use change that has recast ranching from a threat into a key ally for conservation. Although large areas of the region are undeveloped, uncultivated, and un-fragmented, private lands are a small part of the whole, and they are generally located where water and fertile soils are concentrated (Scott et al. 2001). With rapid economic and population growth after World War II, these private lands rose rapidly in market value, and by the end of the century, they were far more valuable as potential residential homesites than as sources of livestock forage (Theobald 2001; Hansen et al. 20022005). As ranches were sold and converted for development, key wildlife habitat was lost in ways that were effectively permanent and greatly exceeded the harm done by livestock in recent decades (Theobald et al. 1997; Brown et al. 2005). Keeping ranches in business began to look like an important strategy for regional conservation of wildlife and open space, especially when compared to the high cost of buying land for public or private preserves.

In two key respects, however, the regulation of wildlife under the Endangered Species Act in the Southwest remains contradictory in ways that adversely affect pastoralism. First, the definition of take contains (or ignores) a contradiction between impacts to *individual specimens* and impacts to*populations* of a listed species. Second, what constitutes an *action* that can be restricted under the Act is ecologically contradictory once one recognizes the degree of long-term human engineering of the region's landscapes. The issue of fire serves to illustrate both contradictions.

More than a century of fire suppression has profoundly altered the Southwest's grasslands and forests, triggering woody encroachment in the former and allowing the buildup of extraordinary fuel loads in the latter. Neither change will undo itself 'naturally' by the removal of its historical causes: shrublands will persist even without livestock grazing, and dense forests will eventually burn at intensities that will result in nearly total tree mortality. The management activities needed to rectify these situations—mechanical removal of shrubs and trees, and/or prescribed fire—may well result in take of *specimens*and, therefore, must meet stringent standards under the Endangered Species Act. But the decision to do nothing, which does not count as an action and, therefore, cannot constitute take under the Act, may threaten listed populations either abruptly (when the inevitable but unpredictable wild fire burns through the overgrown forest) or through the slower, incremental process of grassland-to-shrubland conversion. These contradictions apply to other longstanding regional environmental problems, such as arroyo formation, as well.

The contradictions are most acute on federally owned lands because federal agencies must consult with the Fish and Wildlife Service whenever an action might jeopardize a listed species. Paradoxically, the Forest Service has never had to subject its policy of fire suppression to the process of Endangered Species Act consultation even though it has arguably been the single most important driver of environmental change in the Southwest (and elsewhere in the USA) and can be viewed as ecological engineering in its own right. By contrast, every 10 years, when a Forest Service grazing permit expires, renewal of the permit constitutes an action that must clear the scrutiny of the Act as part of the public consultation process. Given the historical longevity of grazing in the region, however, and the extent to which the landscape has been altered by human activities (for both better and worse), *removing* livestock might constitute a more dramatic ecological act than not doing so (Sayre 2005).

SEPARATING WILDLIFE AND LIVESTOCK IN THE INTERMOUNTAIN WEST

The Intermountain West is an arid region of North America lying between the Rocky Mountains to the east and the Cascades and Sierra Nevada to the west. Cattle ranching and sheep grazing are the most common activities on the uplands in this cold desert steppe (Figure 4), and it is mostly in government ownership with the Forest Service, Bureau of Land Management, and Department of Defense as the primary administrators. Irrigated lands near water are used for agriculture, including hay production, and are often in private ownership. In the southern zones of the region, the winters are short, and there is little precipitation. In the northern portions, the winters are cold and can have significant snow, with hot and dry summers. The vegetation ranges from sparse grasslands and sagebrush to other plants that are able to survive in desert or semidesert conditions in basin lands. Some higher elevation areas have isolated stands of pine, cedar, and aspen tree forests. Idaho is one of the states in this region, and 63% of it is in government ownership (mostly federal), including the largest percentage of Forest Service lands as a percentage of total area within a single state. Bighorn sheep inhabit the montane parts of Intermountain rangelands as do ranchers and shepherds for summer feed for their stock often as part of a pattern of transhumance.

Figure 4: Livestock are moved into the mountains in summer in the Intermountain cold desert steppe. In the mountains, summer range may overlap with habitat for the bighorn sheep.

Agency management to maintain viable populations of wildlife species has become one of the most poignant issues affecting pastoralism and wildlife in the Intermountain West today. Often a function of appeals to regulations designed to implement agency management plans and/or revisions, agency officials now suffer barriers such as a 'culture of fear' from litigation and socio-political conflict associated with contemporary pastoralism and grazing (Knight and White 2009). This cultural and political context not only saddles managers' day-to-day activities but, in effect, presents a significant risk to ranching operations, many of whom became the default stewards of vast working landscapes, open rangelands, and ecological connectivity and services. The recent landmark case of separating domestic and bighorn sheep on the Payette National Forest in Idaho exemplifies the complexities of this trend. In the Payette, agency protocols reflect short-term temporal scales in the move toward separation of species as well as single species management rather than an understanding of the rangelands as a system of interactive effects.

Species Viability and Wildlife Separation

The following recitals appear at the front end of the Idaho Department of Fish and Game (2008)interim policy to manage for separation between bighorn sheep and domestic sheep:

Whereas, bighorn sheep are native to Idaho and have been restored to some historic habitat in partnership with federal land management agencies, other state agencies and private organizations;

Whereas, bighorn sheep are historically, culturally and economically valuable to the citizens of the State of Idaho and Tribes;

Whereas, the domestic sheep industry is historically, culturally and economically valuable to the citizens of the State of Idaho;

Whereas, it is in the public interest to maintain and restore bighorn populations, where appropriate;

Whereas, it is in the public interest to maintain and restore access to state and federal land for grazing by domestic sheep, where appropriate;...

Die-offs from various respiratory-related diseases have chronically affected bighorn sheep populations in the western USA since the mid-1800s (Schommer and Woolever 2008). While bighorn sheep die-offs have occurred in every western state (Toweill and Geist 1999), recent conflicts in Intermountain states over the origins of disease effects on bighorn populations have erupted within the contexts of litigation over public lands grazing, Endangered Species Act dynamics, and the restoration discourse all noted above. In the past 5 years, under these intensifying management and policy conditions, federal and state land management agencies have now developed strategic policies and management protocols that will require species *separation* as a means to ensure minimum viable populations for given areas. Although not currently constrained as a listed species within the Endangered Species Act, the Rocky Mountain subspecies of bighorn sheep holds its status as an iconic and symbolic wildlife species caught within the conflict over land use and management. The implications run deep.

A strong scientific debate continues as to whether populations of domestic sheep (*Ovis aries*) and goats (*Capra hircus*) make bighorn sheep (*Ovis Canadensis*) more susceptible to respiratory diseases and pneumonia. Scientific literature exists along the continuum of the debate, e.g., both disease transmission (Foreyt et al. 1994; Onderka et al. 1988) and the reasons for bighorns' apparent greater susceptibility to *Pasteurella haemolytica* (now equivalent to *Mannheimia haemolytica*) remain unknown and unproven (Dassanayake et al. 2011; Miller et al.

2008). Thus, while most involved in studying and managing the issue readily admit that disease transmission *can occur*, many also recognize that a particular bacterium is not the cause of *all* bighorn die-offs. In fact, a variety of factors other than just disease affect and compound herd health (nutrition, weather, density, predation, management, and stress). Habitat management for bighorn, like many other species, can vary by water availability, noxious weed encroachment, recreation and other disturbance, subspecies migration and transplantation, and landscape connectivity (Idaho Department of Fish and Game 2010). Ongoing research has also continued to test whether vaccinations are an option to avoid the management extremes of separation.

Further, Intermountain rangelands have been severely affected by the invasion of a nonnative annual grass, cheatgrass (*Bromus tectorum*). Cheatgrass, introduced in the nineteenth century in wheat seeds, is a powerful agent of ecosystem engineering. Evidence is strong that it can invade the sagebrush grasslands without human-caused disturbance (Goodrich and Gale 1999), and once it has invaded, it increases the fire frequency by providing a continuous layer of fine fuel. This, in turn, leads to the extirpation of shrubs and herbaceous species that are part of the bighorn diet (Wagner and Peek 2006). The grass itself provides some nutritious green forage for a few months before turning brown and dropping most of its nutritional value for most of the year. The diversity and quality of the diet of grazing animals is severely reduced once the shift to a cheatgrass-dominated state occurs. This has been identified as a potential stressor on bighorn populations (Wagner and Peek 2006) that might contribute to lowered disease resistance and higher mortality from a variety of sources. As in the California case, though livestock have contributed to the spread of the plant, livestock grazing can also be used to reduce the fine fuels and help prevent the cheatgrass-fueled fires that convert the vegetation to cheatgrass dominance. In line with the restoration discourse already described, politically popular approaches to controlling cheatgrass call for seedings with native grasses and shrubs. Unfortunately, the non-equilibrium dynamics of these arid rangelands leads to a high failure rate for such seedings, and cheatgrass, as noted above, invades even intact native grasslands as most native species are not competitive with it. On the other hand, a nonnative perennial grass species, crested wheatgrass (*Agropyron cristatum*), can resist cheatgrass invasion but is politically unpopular because it is nonnative and cannot be considered a valid part of restoration as envisioned in restoration discourse.

For the wool and lamb industry, pastoral uses of the same Intermountain rangeland landscapes traditionally inhabited by bighorn sheep began in the 1840s with domestic sheep crossing southern Idaho along the Oregon Trail. In 1890, the domestic sheep population had grown to over 600,000 and close to three million at the end of World War I (Hart 2009). Scottish and Basque family heritage have comprised much of the multi-generational industry sustained today in Idaho, with total herd numbers reduced to between 220,000 and 235,000 (National Agricultural Statistics Service 2011). Despite the substantive change in domestic sheep production in Idaho, the industry has experienced continued demand for lamb and wool, especially recently in more globalized markets.

Nearly half of all domestic sheep produced in the USA spend time grazing on public land allotments. Although 80% of sheep producers are east of the Mississippi River, a full 80% of the sheep exist within operations west of the Mississippi River, making the public-private lands structure of many larger operations more striking with regard to overall industry production. To put this scale in perspective though, in Idaho, the Forest Service manages roughly 22 million acres including about 5.3 million acres of bighorn habitat. Domestic sheep allotments overlap a small percentage of the bighorn habitat, confined to 138,000 acres; however, this overlap constitutes critical summer forage range within the pastoral migration patterns. Culturally and economically, the bighorn represents multiple values in the state of Idaho for wildlife enthusiasts and hunters, the latter of which generate significant revenue for the state management agency through hunting license and tag sales. The domestic sheep industry also sustains a long-standing and renowned annual 'trailing of the sheep' to celebrate the traditional migration of sheep bands to winter range elevations coordinated with a fall cultural and heritage festival. Currently, almost all sheepherders in Idaho are native Peruvians as willing and skilled American sheepherders have disappeared from the Intermountain labor force.

Turning point of the 'Payette principles'

In 2003, Idaho's Payette National Forest revised its management plan. Concerns about whether or not it allowed adequate viability of bighorn populations led to an appeal eventually upheld by the chief of the Forest Service in 2005. The following year, the Forest Service produced a risk analysis report to address bighorn viability on the forest and convened a panel of 11 scientific experts resulting in a consensus of the 'Payette Principles' (Payette National Forest Science Panel 2006). The principles are broad and suggest a great deal of uncertainty about disease transmission between the wildlife and livestock but perhaps more importantly also planted the seed for 'separation policy' as a means to address the viability mandates. Regulations regarding implementation have continued to evolve and remain controversial over the last 30 years, including the definition and use of what constitutes 'viable populations' for species the Forest Service is required to sustain.

The emergence of new demands on the 'Minimum Viable Population rule' eventually led to significant management change within the Payette National Forest. In 2010, the Forest Service issued a Record of Decision to cancel 70% of the grazing allotments on the Payette National Forest in order to achieve separation of bighorns and domestic sheep. Due to the inability to prove that disease transmission would not occur if the Forest Service maintained the leases, the agency could not guarantee a minimum viable population for the bighorn and, thus, began canceling grazing permits for allotments in the Payette with the plan to phase in additional cancellations. Ambiguity remains as to whether all affected allotments constitute habitat and range for bighorns. The agency has also required additional on-the-ground management strategies (e.g., daily global positioning system records of each domestic herd).

Because the Payette National Forest lies within the state boundaries of Idaho, and bighorn are managed by the Idaho Department of Fish and Game, the state also 'acted' amidst the turmoil of the management plan debate and prior to the 2010 Forest Service decision. Idaho governor Butch Otter convened a collaborative task force, including scientists, industry representatives, wildlife advocates, political representatives, tribal representatives, and agency officials. The diverse group originally agreed to the following mission statement: *To maintain and restore bighorn sheep populations without causing undue economic hardship on either the domestic sheep industry in Idaho or individual Idaho sheep producers* (Idaho Department of Fish and Game 2010, page 2). Issues of federalism crept into the States' 2009 legislative session, however, with multiple bills proposed to address the issue:

SB1175 proposed the state kill all bighorn sheep that came into contact with domestic sheep was passed by both the State Senate and House and, then, vetoed by Governor Otter; subsequently,

SB1232a proposed the Idaho Department of Fish and Game, in collaboration with the woolgrowers, develop best management practices for domestic sheep operations on the Payette National Forest with the *intent of separation* but that would give assurances to producers as well as Forest Service personnel about maintaining viable populations of bighorns.

SB1232a only required voluntary participation by sheep producers (eventually 18 of 21 agreements were made for the Payette) and was signed into law but presented challenges in that it was not clear whether the state of Idaho or the Forest Service had ultimate control over the outcomes for producers with respect to management and policy for wildlife-livestock interactions. The conflict intensified such that several participating environmental groups, along with the Nez Perce Indian Tribe, withdrew from the state-level advisory group attempting to provide collaborative recommendations. While both the federal- and state-level actions represent macro-strategies toward engineering the landscape, the recent actions illustrate the variability in approaches at the science-policy interface affecting vast areas of rangelands in the West. The Payette case also demonstrates that 'disturbance' with effect(s) to the landscape can occur through the courts, routine and required public consultation processes, and the rhetoric that accompanies management debates. Yet, rarely are such social disturbances accounted for in agency policy decision making.

Restoration and precedence

While no definitive agreement exists, estimated bighorn sheep populations were higher historically, suggesting an overall decline despite habitat and range having remained the same. In some areas where bighorn have been transplanted, herds have re-established with varying success. Most states do not re-establish bighorn proximate to domestic sheep allotments (Schommer and Woolever 2008); however, in 1997, after a series of transplanted populations were reintroduced to the Hells Canyon region, the Forest Service, Idaho Department of Fish and Game, and the Hells Canyon Bighorn Sheep Restoration Committee made an agreement with producers grazing in Hells Canyon that stated 'the potential risk, if any, of disease transmission and loss of bighorn sheep when the same invade domestic livestock or sheep operations is accepted' (O'Laughlin and Cook 2010).

No agreed to level of viable populations (minimum, or not) exists among the various stakeholders for situations like the bighorn/domestic sheep conflict. Although some have recognized resource management as a social problem for decades (Schroeder, no date), the volatility of the conflict and strategies now used to affect and influence management and agency actions charge the overall management context with a type of dilemma for which many resource managers remain untrained. Social scientists might point out that the situation reveals that at least some stakeholders have vacated the *social contract* required for resource management to translate to environmental sustainability, stewardship, and ecological integrity. Some argue that paradoxically, the way environmental laws are 'used' to litigate against activities such as grazing in today's pastoral landscapes not only miss the mark of the original laws' intents but also, in some cases, could *prevent*the very environmental benefits the laws have as their core objectives (Nie 2008). The lack of a social license of cooperation, so to speak, that would bind stakeholders together (friends and foes alike) indicates the landscape, rural communities, and many of the resources that sectors of the population take for granted will continue to remain at risk, perhaps increasingly so. At least at the intersection of public lands and wildlife-livestock interactions, a 'congress' is needed to reinvigorate a spirit of cooperation to enable more sustainable change.

The pattern of far-reaching agency decisions to implement new policies such as with the Payette National Forest's Record of Decision for the bighorn sheep is also at risk of having ripple effects. In Idaho, although not mandated to do so, the Bureau of Land Management has already mapped 30-mile 'circles' around known bighorn sheep populations on lands it manages with the intent to eliminate all domestic sheep and goats in those ranges. Given Idaho's landscape and very urban population in a still very rural landscape, this will not only further affect domestic commercial sheep producers but also more mundane domestic contexts, such as families and children in farm youth programs learning pastoral skills and animal husbandry with livestock. More far reaching, the precedents set by the Payette decision could also 'jump' to begin to affect other species involved in other wildlife-livestock interactions in other regions (e.g., bison and cattle in Montana). These decisions may also have profound long term effects on the landscape by removing livestock as a potential management tool in attempting to stabilize the state change the rangelands are undergoing as a result of cheatgrass invasion. A failure to prevent this change is a serious long-term threat to the bighorn. In the Payette case, policy did not emerge absent from biological assessment but uses separation as a tool to articulate a biological stance within the working landscape.

CONCLUSIONS

Ranchers in the USA can act as crucial agents for wildlife conservation. Yet, in the discourse of wildlife conservation and restoration, they are persistently tagged as suspects in a narrative of environmental degradation. Regulations and policies often institutionalize this positioning, reflected in a story told by a rancher to students on a field trip in 2009:

The wildlife biologist brought what looked like a shoebox filled with tiger salamanders out to my ranch for relocation. Her family was from India, she was educated in the city, and she had never been on anything like my ranch before. I drove her out to the pond and she took the salamanders out and they seemed happy to find themselves at the pond. We had to be careful—since I am not an approved wildlife technician, I am not allowed to touch them.

Wildlife policy and a shift to a greater emphasis on biodiversity have had both positive and negative impacts on pastoralists. Where pastoralists own the habitat, regulations are a burden, but offering incentives can make having a rare species an opportunity instead of a risk. However, these incentives are not supported unless the benefits of maintaining grazing are recognized. Even so, the relationship between grazing as a local process and the need to maintain enterprises, communities, and landscape configurations is seldom appreciated. In the mixed ownerships of the southwest, alliances linking public and private lands have helped to stabilize land tenure and maintain habitat. Yet, wildlife and environmental policies are ill-equipped to resolve contradictions between conservation of individuals versus populations, or fire suppression and grazing removal as inactions that require no review versus grazing and burning as actions that require regulation and control. In the Intermountain West, where the majority of range is in federal management, removal of stock to enhance wildlife is not uncommon even though knowledge of wildlife-livestock interactions remains limited. The impacts of cessation of grazing on other species or of the loss of ranching enterprises on long-term conservation of unfragmented rangelands are not considered.

Throughout the western USA, pastoralism-wildlife interactions are taking place in a changed and changing environment that can, in some cases, be understood as ecosystem engineering. The long-term change scenarios posited by Perevolotsky and Seligman (1998) for the Mediterranean Basin offer potential insight here. They suggest that the introduction of livestock grazing leads to a state change in ecosystems that eventually 'stabilizes' around a new set of state characteristics that are maintained by grazing, but not reversed by a removal of grazing. In fact, livestock removal can lead to another state change and a reduction in biodiversity. To some degree, the three cases presented here might fit this model, though the relative time span in the USA is quite short. However, what is clear in all the cases is that a focus on 'restoration' is unrealistic, and the failure to consider removal of livestock grazing as an ecologically potent action may in fact lead to further and more unfortunate ecosystem change. Absent adequate ecological information, the assumptions of innate conflict between livestock and wildlife, and that cessation of grazing is not an action, as well as the norms of a politically popular yet ecologically unsupportable discourse of restoration fill in the gaps. It is unfortunate that the available site-specific ecological information is most often quite limited.

Adaptive management is now the go to solution for situations where management takes place in a setting of 'ecological uncertainty'. It is clear that the adaptive management process can help improve the scientific information for a site, and that it encourages the flexible and responsive management that changing and non-equilibrial rangelands demand. Worth investigation is how adaptive management can and cannot work to help overcome the assumptions and norms that cloud management decisions and that obscure the linkages across local to landscape scales.

Pastoralism is in decline in the USA. As we learn more about the roles of ranching and grazing in creating and maintaining habitat at local and landscape scales, it behooves us to be concerned that the 'supply' of pastoralists is in danger (Brunson and Huntsinger 2008). Once removed, it may be that returning grazing to an ecosystem is impossible or prohibitive. Removing grazing from government lands may result in a permanent loss of valuable range and habitat, as ranches are sold and lands are developed as part of a westwide expansion of second homes and residential development (Sulak and Huntsinger 2007). Maintaining grazing on government lands without consideration of planning and protection for private lands may leave ranchers literally without a 'home'.

In all cases, the existence of bureaucratic systems of regulation and enforcement is perplexing to pastoralists. A common observation is that the government does things the most expensive way possible. Another is that ranchers are being punished for historically maintaining wildlife habitat as having that wildlife on public or private lands winds up reducing access and increasing costs. Not only is this a confusing message to send to food producers but, as landscape managers affecting ecological services for society as a whole, reduced access and increased cost signal the wrong incentives to them as stewards. For ranchers, an innate enjoyment of the environment and nature extends to wildlife, and pastoralists have shown themselves to be ready to steward rare species. Of the diverse groups attempting to conserve the landscape, many are increasingly seeing the multi-scale benefits–landscape and local–of pastoral systems despite institutional and legal barriers. These linkages must be fostered and maintained if extensive rangeland grazing is to persist in the USA.

AUTHORS' CONTRIBUTIONS

LH conceived of and led the paper, drafted the introduction, background, and conclusions, and wrote the first case study. NS contributed to the introduction, background, and conclusions, and wrote the second case study. JW wrote the third case study. All authors read and approved the final manuscript.

REFERENCES

1. Alameda County. 2011. Wildlife friendly ponds program. http://www.acrcd.org/ForRuralLandowners/WildlifeFriendlyPondsProgram.aspx Accessed Oct. 30, 2011
2. Bahre, CJ, and ML Shelton. 1996. Rangeland destruction: Cattle and drought in southeastern Arizona at the turn of the century. Journal of the Southwest 38: 1–22.
3. Balestreri, AN. 1981. Status of the San Joaquin kit fox at Camp Roberts, California. Unpubl. Report, US Department Army, Environmental, and Natural Resources Office.. San Luis Obispo: California Polytechnic State University.
4. Bartolome, JW, and RD Jackson. 2007. Grazing ecology of California grasslands. In Ecology and Management of California Grasslands, ed. Jeffrey C, Mark S, D'Antonio CM, Berkeley: UC Press.
5. Barry, SJ. 2011. Current findings on grazing impacts: California's special status species benefit from grazing. California Cattlemen's Association Magazine. June:18–20. http://www.carangeland.org/images/11-06_CCA_Mag_-_Current_Findings_on_Grazing_Impacts.pdf. Accessed 11 March 2012.
6. Behnke, RH, I Scoones, and C Kerven. 1993. Range ecology at disequilibrium: new models of natural variability and pastoral adaptation in African savannas. London: Overseas Development Institute.
7. Blench, R. 2001. You can't go home again: Pastoralists in the new millennium. Animal Production and Health Paper. Rome: Food and Agriculture Organization.http://www.odi.org.uk/work/projects/pdn/eps.pdf Accessed Oct. 30, 2011
8. Brown, DG, KM Johnson, TR Loveland, and DM Theobald. 2005. Rural land-use trends in the coterminous United States, 1950–2000. Ecological Applications 15: 1851–1863
9. Brown, JH, and A Kodric-Brown. 1996. Biodiversity on the borderlands. Natural History 105: 58–61.
10. Brunson, M, and L Huntsinger. 2008. Can old ranchers save the new west? Synthesis paper. Journal of Range Ecology and Management 61: 137–147.

11. Cooke, RU, and RW Reeves. 1976. Arroyos and environmental change in the American southwest. Oxford: Clarendon Press.
12. California Rangeland Trust. 2011. About the California Rangeland Trust. http://www.rangelandtrust.org/about_us.php Accessed Oct. 31, 2011
13. Dassanayake, RP, KP Paulraj Jr, DP Knowles, WC Davis, WJ Foreyt, and S Srikumaran. 2011. Co-expression of ovine LPS receptor CD14 with Mannheimia haemolytica leukotoxin receptor LFA-1 or Mac-1 does not enhance leukotoxin-induced cytotoxicity. Veterinary Immunology and Immunopathology 14 (1–2): 84–91.
14. Dennis, P, MR Young, CL Howard, and IJ Gordon. 1997. The response of epigeal beetles (*Col. Carabidae, Staphylinidae*) to varied grazing regimes on upland *Nardus stricta* grasslands. Journal of Applied Ecology 34: 433–443. Publisher Full Text
15. DiDonato, J. 2007. Endangered amphibian research within grazed grasslands. In Keeping Landscapes Working: A University California Cooperative Extension Newsletter for Rangeland Managers. Winter:4–6. http://cesantaclara.ucdavis.edu/newsletterfiles/newsletter710.htm *webcite* Accessed 5 October 2007
16. Dobson, Andrew P, J Rodriguez, W Roberts, and D Wilcove. 1997. Geographic distribution of endangered species in the United States. Science 275: 550
17. Dobyns, HF. 1981. From fire to flood: Historic human destruction of Sonoran desert riverine oases. Anthropological Papers No. 20. Socorro, NM, USA: Ballena Press.
18. Environmental Defense Fund. 2007. Alameda rancher and wildlife-friendly program restores ponds and helps rare animals. Center for Conservation Incentives October 10. http://www.edf.org/article.cfm?contentID=6295 Accessed Oct 30 2011
19. Ferranto, S, L Huntsinger, C Getz, G Nakamura, W Stewart, S Drill, Y Valachovic, M DeLasaux, and M Kelly. 2011. Forest and rangeland owners value land for natural amenities and as financial investment. California Agriculture 65 (4): 185–191.
20. Ferranto, S, H Lynn, N Gary, S William, G Christy, and K Maggi. In press. Consider the source: The impact of media and authority in outreach to private forest and rangeland owners. Journal of Environmental Management.

21. Foreyt, WJ, KP Snipes, and RW Kasten. 1994. Fatal pneumonia following inoculation of healthy bighorn sheep with Pasteurella haemolytica from healthy domestic sheep. Journal of Wildlife Diseases 30 (2): 137–145
22. Goodrich, S, and G Natalie. 1999. In Proceedings: ecology and management of pinyon-juniper communities within the Interior West; 1997 September 15–18; Provo, UT. Proc. RMRS-P-9. Ogden, UT: U.S. Department of Agriculture, Forest Service, Rocky Mountain Research Station, ed. Monsen S, Stephen B, Stevens S, Richard S, http://www.fs.fed.us/rm/pubs/rmrs_p009/rmrs_p009_069_071.pdf *webcite* Accessed Dec. 26, 2011
23. Gosnell, H, and WR Travis. 2005. Ranchland ownership dynamics in the Rocky Mountain West. Rangeland Ecology & Management 58 (2): 191–198.
24. Hansen, AJ, RL Knight, JM Marzluff, S Powell, K Brown, PH Gude, and K Jones. 2005. Effects of exurban development on biodiversity: patterns, mechanisms, and research needs. Ecological Applications 15: 1893–1905.
25. Hansen, AJ, R Rasker, et al.. 2002. Ecological causes and consequences of demographic change in the New West. BioScience 52: 151–162
26. Hart, AA. 2009. Idaho's rangelands: a history. Caxton Printers, Caldwell, ID.
27. Heady, HF. 1958. Vegetational changes in the California annual type. Ecology 39 (3): 402–416.
28. Heady, HF. 1977. Valley grassland. In Terrestrial vegetation of California, ed. Barbour MG, Major J, California Native Plant Society, Sacramento, CA.
29. Herbel, CH, and RP Gibbens. 1996. Post-drought vegetation dynamics on arid rangelands of southern New Mexico.
30. Huntsinger, L, JW Bartolome, and CM D'Antonio. 2007. Grazing management of California grasslands. In In ecology and management of California grasslands, ed. Corbin J, Stromberg M, D'Antonio CM, Univ. Ca. Press, Berkeley, CA.
31. Huntsinger, L, LC Forero, and A Sulak. 2010. Transhumance and pastoralist resilience in the western United States. Pastoralism: Research, Policy, and Practice 1 (1): 1–15.

32. Idaho Department of Fish and Game. 2008. Interim strategy for managing separation between bighorn sheep and domestic sheep in Idaho. Boise, ID.
33. Idaho Department of Fish and Game. 2009. Idaho bighorn sheep species management plan. Presentation to the Idaho Bighorn / Domestic Sheep Advisory Group. Boise, ID.
34. Idaho Department of Fish and Game. 2010. Bighorn Sheep / Domestic Sheep Advisory Group: Recommendations and final report. Idaho Department of Fish and Game and Idaho Department of Agriculture, Boise, ID.
35. Kaib, M, CH Baisan, HD Grissino-Mayer, TW Swetnam, et al.. 1996. Fire history in the gallery pine-oak forests and adjacent grasslands of the Chiricahua mountains of Arizona. In tech. coord. In Effects of fire on Madrean Province ecosystems: A symposium proceedings. USDA Forest Service General Technical Report RM-GTR-289, ed. Peter-Ffolliott F, 253–264. Rocky Mountain Forest and Range Experiment Station, Fort Collins, CO.
36. Kelt, DA, ES Konno, and JA Wilson. 2005. Habitat management for the endangered Stephens' kangaroo rat: The effect of mowing and grazing. Journal of Wildlife Management 69 (1): 424–429
37. 2009. Conservation for a New Generation: Redefining Natural Resource Management,Knight RL, White C Island Press, Washington, D.C.
38. Kreuter, UP, MV Nair, D Jackson-Smith, JR Conner, and JE Johnston. 2006. Property rights orientations and rangeland management objectives: Texas, Utah, and Colorado. Rangeland Ecology & Management 59 (6): 632–639.
39. Lee, , and K Henry. 2001. Probation given for relocating endangered frogs / Consultant also fined $10,000. San Francisco Chronicle, February 23, 2001. http://articles.sfgate.com/2001-02-23/news/17586700_1_frogs-calaveras-county-wetlands-society*webcite*Accessed Oct.29, 2011)
40. Lehman, W. 2007. Malpai borderlands Habitat Conservation Plan. AESO/SE 22410-2006-F-0408.
41. Liffmann, RJ, L Huntsinger, and LC Forero. 2000. To ranch or not to ranch: Home on the urban range? Journal of Range Management 53 (4): 362–370.

42. Lueck, D. 2002. The extermination and conservation of the American Bison. The Journal of Legal Studies 31 (2): S609–S652
43. Lueck, D, and JA Michael. 2003. Preemptive habitat destruction under the Endangered Species Act. Journal of Law and Economics 46 (1): 27–60.
44. Maestas, JD, RL Knight, and WC Gilgert. 2003. Biodiversity across a rural land-use gradient. Conservation Biology 17 (5): 1425–1434.
45. Marty, JT. 2005. Effects of cattle grazing on diversity in ephemeral wetlands. Conservation Biology 19 (5): 1626–1632
46. Miller, , W Mike, DP Knowles, and MS Bulgin. 2008. Pasteurellosis transmission risks between domestic and wild sheep. CAST Commentary QTA2008-1. Council for Agricultural Science and Technology, Ames, IA.
47. National Agricultural Statistics Service. 2011. Sheep and goats. http://usda.mannlib.cornell.edu/usda/current/SheeGoat/SheeGoat-01-28-2011.pdf *webcite*Accessed Oct. 30, 2011
48. Niamir-Fuller, M. 1998. The resilience of pastoral herding in Sahelian Africa. In Linking social and ecological systems, ed. Berkes F, Folke C, Cambridge University Press, Cambridge.
49. Nie, M. 2008. The governance of western public lands: mapping its present & future. University Press of Kansas, Lawrence, KS.
50. Nuzum, , and C Robert. 2005. Report: Using livestock grazing as a resource management tool in California. Contra Costa Water District report, Concord, CA. http://www.ccwater.com/files/LivestockGrazingFinal72005.pdf *webcite* Accessed August 9 2006
51. Oakes, and L Claudia. 2000. History and consequence of keystone mammal eradication in the desert grasslands: the Arizona black-tailed prairie dog (Cynomys ludovicianus arizonensis). PhD dissertation. University of Texas, Austin, TX.
52. O'Laughlin, J, and PS Cook. 2010. Bighorn sheep and domestic sheep: current situation in Idaho. Report No. 30, Policy Analysis Group, College of Natural Resources. University of Idaho, Moscow, ID.
53. Onderka, DK, SA Rawluk, and WD Wishart. 1988. Susceptibility of Rocky Mountain bighorn sheep and domestic sheep to

pneumonia induced by bighorn and domestic livestock strains of Pasteurella haemolytica. Canadian Journal of Veterinary Research 52: 439–444

54. Perevolotsky, A, and No'am Seligman. 1998. Role of grazing in Mediterranean rangeland ecosystems-inversion of a paradigm. BioScience 48 (12): 1007–1017.
55. Pyke, CR, and JT Marty. 2005. Cattle grazing mediates climate change impacts on ephemeral wetlands. Conservation Biology 19 (5): 1619–1625
56. Richmond, MW Orien, J Tecklin, and SR Beissinger. fv. In revision. Impact of cattle grazing on the occupancy of a cryptic, threatened rail. Ecological Applications.
57. Richmond, OMW, SK Chen, BB Risk, J Tecklin, and SR Beissinger. 2010. California black rails depend on irrigation-fed wetlands in the Sierra Nevada foothills. California Agriculture 64 (2): 85–93
58. Sayre, NF. 1999. The cattle boom in southern Arizona: Towards a critical political ecology. Journal of the Southwest 41 (2): 239–271.
59. Sayre, NF. 2002. Ranching, endangered species, and urbanization in the southwest: Species of capital. University of Arizona Press, Tucson.
60. Sayre, A, and F Nathan. 30 Octoberâ "1 November 2003.. Recognizing history in range ecology: 100 years of science and management on the Santa Rita Experimental Range. In In Santa Rita Experimental Range: 100 Years (1903–2003) of accomplishments and contributions, ed. McClaran MP, Ffolliott PF, Edminster CB, 1–15. USDA Forest Service Proceedings RMRS-P-30, U.S. Department of Agriculture, Forest Service, Rocky Mountain Research Station, Ogden, UT, USA,.
61. Sayre, NF. 2005. Interacting effects of landownership, land use, and endangered species on conservation of southwestern U.S. rangelands. Conservation Biology 19 (3): 783–792.
62. Sayre, NF. 2010. Climax and 'original capacity': The science and aesthetics of ecological restoration in the southwestern USA. Ecological Restoration 28: 23–31.
63. Schommer, TJ, and MM Woolever. 2008. A Review of Disease Related Conflicts Between Domestic Sheep and Goats and

Bighorn Sheep. United States Department of Agriculture Forest Service, Rocky Mountain Research Station: General Technical Report RMRS-GTR-209. Fort Collins, CO, USA.

64. Schroeder, W. No date. Management by Court Decree. Miscellaneous Publication #4, College of Forestry, Range, and Wildlife Sciences. University of Idaho, Moscow, ID,USA.
65. Scott, T, R Standiford, and N Pratini. 1995. Private landowners critical to saving California biodiversity. California Agriculture 49: 50–57. Publisher Full Text
66. Scott, JM, FW Davis, RG McGhie, RG Wright, C Grover, and J Estes. 2001. Nature reserves: Do they capture the full range of America's biological diversity? Ecological Applications 11: 999–1007
67. Sulak, A, and L Huntsinger. 2007. Public lands grazing in California: untapped conservation potential for private lands? Rangelands 23 (3): 9–13.
68. Swetnam, TW, CD Allen, and JL Betancourt. 1999. Applied historical ecology: Using the past to manage for the future. Ecological Applications 9: 1189–1206.
69. Swetnam, TW, and JL Betancourt. 1998. Mesoscale disturbance and ecological response to decadal climatic variability in the American southwest. Journal of Climate 11: 3128–3147.
70. Theobald, DM. 2001. Land-use dynamics beyond the American urban fringe. Geographical Review 91: 544–564
71. Theobald, DM, JR Miller, and NT Hobbs. 1997. Estimating the cumulative effects of development on wildlife habitat. Landscape and Urban Planning 39: 25–36.
72. Toweill, DE, and V Geist. 1999. Return to royalty: wild sheep of North America. Boone and Crockett Club and Foundation for North American Wild Sheep, Missoula, MT.
73. Turner, RM, RH Webb, JE Bowers, and JR Hastings. 2003. The changing mile revisited: an ecological study of vegetation change with time in the lower mile of an arid and semiarid region. University of Arizona Press, Tucson, AZ.
74. United States Code. 16 U.S.C. ÂÂ§ 1604(g)(3)(B).

75. 2011. USDA-FS [United States Department of Agriculture Forest Service]. About us. http://www.fs.fed.us/aboutus/ Accessed Oct. 30, 2011

76. USDI-BLM [United States Department of Interior Bureau of Land Management]. 2011. About the BLM. http://www.blm.gov/wo/st/en/info/About_BLM.html *webcite* Accessed Oct. 30, 2011

77. USDI-FWS [United States Department of Interior Fish and Wildlife Service]. 1997. Draft recovery plan for the Stephen's kangaroo rat, Region 1. United States Department of Interior Fish and Wildlife Service, Portland, Oregon. http://ecos.fws.gov/docs/recovery_plan/970623.pdf *webcite* Accessed Oct. 30, 2011

78. USDI-FWS [United States Department of Interior Fish and Wildlife Service]. 2002. Endangered and Threatened wildlife and plants: Listing of the Chiricahua Leopard Frog (Rana chiricahuensis). Federal Register 67: 40790–40811.

79. USDI-FWS [United States Department of Interior Fish and Wildlife]. 2004. Endangered and threatened wildlife and plants; Determination of threatened status for the California Tiger Salamander; and Special Rule Exemption for existing routine ranching activities; Final Rule. Federal Register 69, No. 149, 50 CFR Part 17, RIN 1018-A168, 69 FR 68568.

80. USDI-FWS [United States Department of the Interior-Fish and Wildlife Service]. 2006. Designation of critical habitat for the California Red-Legged Frog, and Special Rule Exemption associated with final listing for existing routine ranching activities; Final Rule. Fish and Wildlife Service, Federal Register 71 FR 19244 19346.

81. USDI-FWS. [U. S. Department of Interior Fish and Wildlife Service]. 2011. ESA Basics: More than 30Â years of conserving endangered species. U. S. Department of Interior Fish and Wildlife Service Endangered Species Program, Arlington, VA.

82. USDI-FWS [U. S. Department of Interior Fish and Wildlife Service]. 2002. Endangered and Threatened Wildlife and Plants; Listing of the Chiricahua Leopard Frog (Rana chiricahuensis); Final Rule Federal Register 67 FR 40790 40811.

83. Weiss, SB. 1999. Cars, cows, and checkerspot butterflies: Nitrogen deposition and management of nutrient-poor grasslands for a threatened species. Conservation Biology 13 (6): 1476–1486

84. Westoby, M. 1980. Elements of a theory of vegetation dynamics in arid rangelands. Israel Journal of Botany 28: 169–194.

85. Wagner, Guy D, and James M Peek. 2006. Bighorn sheep diet selection and forage quality in central Idaho. Northwest Science 80 (4): 246–258.

86. Weiss, Stuart B. 1999. Cars, cows, and checkerspot butterflies: Nitrogen deposition and management of nutrient-poor grasslands for a threatened species. Conservation Biology 13 (6): 1476–1486

87. Westoby, Mark. 1980. Elements of a theory of vegetation dynamics in arid rangelands. Israel Journal of Botany 28: 169–194.

Citations

CHAPTER 1

Abdel Wahab, D. (2014), Introduction of Fertigation in Sugarcane Production for Optimization of Water and Fertilizers Use. Agricultural Sciences, 5, 945-957, doi: 10.4236/as.2014.511102.

CHAPTER 2

Karin Fazeni and Horst Steinmüller, Impact of changes in diet on the availability of land, energy demand, and greenhouse gas emissions of agriculture, doi: 10.1186/2192-0567-1-6.

CHAPTER 3

Patricia Möller and Andrés Muñoz-Pedreros, Legal Protection Assessment of Different Inland Wetlands in Chile, doi:10.1186/s40693-014-0023-1.

CHAPTER 4

Jennifer Firn, Jodi N Price, and Ralph DB Whalley, Using Strategically Applied Grazing to Manage Invasive Alien Plants in Novel Grasslands, doi:10.1186/2192-1709-2-26.

CHAPTER 5

Marion Dolezel, Marianne Miklau, Angelika Hilbeck, Mathias Otto, Michael Eckerstorfer, Andreas Heissenberger, Beatrix Tappeser, and Helmut Gaugitsch, Scrutinizing the current practice of the environmental risk assessment of GM maize applications for cultivation in the EU, doi: 10.1186/2190-4715-23-33.

CHAPTER 6

Kristin Monie, Singarayer Florentine, and Grant Palmer, Recruitment and Functionality Traits as Bioindicators of Ecological Restoration Success in the Lurg Hills District, Victoria, Australia, doi:10.1186/2192-1709-2-27.

CHAPTER 7

Chenicheri Chandroth Sreejith, Chandrasekharan Muraleedharan, and Palatel Arun, Life Cycle Assessment of Producer Gas Derived from Coconut Shell and its Comparison with Coal Gas: An Indian Perspective, doi:10.1186/2251-6832-4-8.

CHAPTER 8

Lynn Huntsinger, Nathan F Sayre, and JD Wulfhorst, Birds, Beasts and Bovines: Three Cases of Pastoralism and Wildlife in the Usa, doi: 10.1186/2041-7136-2-12.

Index

G

I

L

M

N

O

P

R

S

W